CW00675425

Synthetic Philosophy of Contemporary Mathematics

FERNANDO ZALAMEA

Synthetic Philosophy of Contemporary Mathematics

Translated by

ZACHARY LUKE FRASER

URBANOMIC

sequence

Published in 2012 by

URBANOMIC
THE OLD LEMONADE FACTORY
WINDSOR QUARRY
FALMOUTH TR11 3EX
UNITED KINGDOM

SEQUENCE PRESS
36 ORCHARD STREET
NEW YORK
NY 10002
UNITED STATES

Originally published in Spanish as
Filosofía Sintética de las Matemáticas Contemporáneas
© Editorial Universidad Nacional de Colombia, 2009
This translation © Sequence Press

LIBRARY OF CONGRESS CONTROL NUMBER
2012948496

BRITISH LIBRARY CATALOGUING-IN-PUBLICATION DATA

A full record of this book is available
from the British Library

ISBN 978-0-9567750-1-6

Copy editor: Daniel Berchenko
Printed and bound in the UK by
the MPG Books Group, Bodmin and Kings Lynn

www.urbanomic.com
www.sequencepress.com

CONTENTS

All that we call invention, discovery in the highest sense of the word, is the meaningful application and the putting into practice of a very original feeling of truth, which, over a long and secret period of formation, leads unexpectedly, with lightning speed, to some fertile intuition. [...] It is a **synthesis** *of world and spirit that offers the most sublime certainty of the eternal harmony of existence.*

GOETHE, *WILHELM MEISTER'S APPRENTICESHIP* (1829)

Poetry points to the enigmas of nature and aims to resolve them by way of the imagination; **philosophy** *points to the enigmas of reason and tries to resolve them by way of words.*

GOETHE, *POSTHUMOUS FRAGMENTS*

The most important thing, nevertheless, continues to be the **contemporary**, *because it is what most clearly reflects itself in us, and us in it.*

GOETHE, *NOTEBOOKS ON MORPHOLOGY* (1822)

Mathematicians *are a bit like Frenchmen: when something is said to them, they translate it into their own language, and straight away it becomes something else entirely.*

GOETHE, *POSTHUMOUS FRAGMENTS*

Introduction

OPTIONS TRADITIONALLY AVAILABLE TO MATHEMATICAL PHILOSOPHY, AND A PROSPECTUS OF THE ESSAY

Drawing on the four maxims of Goethe's that we have placed as epigraphs, we would like to explain, here, the general focus of this *Synthetic Philosophy of Contemporary Mathematics*. The four terms in the title have, for us, certain well-defined orientations: 'synthetic' points to the connective, relational environment of mathematical creation and to a veiled reality that contrasts with invention; 'philosophy', to the reflective exercise of reason on reason itself; 'contemporary', to the space of knowledge elaborated, broadly speaking, between 1950 and today; 'mathematics', to the broad scope of arithmetical, algebraic, geometrical and topological constructions – going beyond merely logical or set-theoretical registers. Now, in setting the stage in this way, we have immediately indicated what this essay *is not*: it is from the outset clear that it will *not* be a treatise on the 'analytic philosophy of the foundations of mathematics in the first half of the twentieth century'. Since the great majority of works in mathematical philosophy (chapter 2) fall exclusively within the subbranch those quotation marks encapsulate, perhaps we can emphasize the interest that lies in an essay like this one, whose visible spectrum will turn out to be virtually *orthogonal* to the one usually treated in reflections on mathematical thought.

These pages seek to defend four central theses. The first postulates that the conjuncture 'contemporary mathematics' deserves to be investigated with utmost care, and that the modes of doing advanced mathematics cannot be reduced (chapters 1, 3) to either those of set theory and mathematical logic, or those of elementary mathematics. In this investigation, we hope to introduce the reader to a *broad spectrum of mathematical achievements in the contemporary context,* which might have otherwise remained inaccessible. The second thesis says that to really see, even in part, *what is happening* within contemporary mathematics (chapters 4–7), we are practically forced to expand the scope of our vision and discover the *new problematics* at stake, undetected by 'normal' or 'traditional' currents in the philosophy of mathematics (chapters 2–3). The third thesis proposes that a turn toward a *synthetic* understanding of mathematics (chapters 3, 8–11) – one that is largely reinforced in the mathematical theory of categories (chapters 3–7) – allows us to observe important *dialectical tensions* in mathematical activity, which tend to be obscured, and sometimes altogether erased, by the usual analytic understanding. The fourth thesis asserts that we must reestablish a vital *pendular weaving between mathematical creativity and critical reflection* – something that was indispensable for Plato, Leibniz, Pascal and

Peirce – and that, on the one hand, many present-day mathematical constructions afford useful and original perspectives on certain philosophical problematics of the past (chapters 8–11), while, on the other hand, certain fundamental philosophical *insolubilia* fuel *great creative forces* in mathematics (chapters 3–7, 10).

The methods utilized in this work include *describing* a particular state of affairs ('contemporary mathematics': chapters 4–7), *reflecting* on this description ('synthetic philosophy': chapters 1, 3, 8–11), and *contrasting* this two-fold description and reflection with other related aspects ('mathematical philosophy', theory of culture, creativity: chapters 2, 8–11). We hope to make the hypotheses underlying those descriptions, reflections and comparisons explicit throughout. Note that our survey's main endeavor has been to try to observe mathematical movements *on their own terms*, and that its filter of cultural reorganization has been articulated only a posteriori, so as to try to reflect as faithfully as possible those complex, and often elusive, movements of mathematics.

The problems that mathematics have posed for philosophical reflection have always been varied and complex. Since the beginnings of both disciplines in the Greek world, the advances of mathematical technique have perpetually provoked philosophical reflections of a fundamental nature. The privileged *frontier* of mathematics – the fluctuating, intermediary warp between the possible (hypothesis), the actual (comparisons) and the necessary (demonstrations), the bridge between human inventiveness and a real, independent world – has spawned all kinds of alternative positions regarding what mathematics 'is', what its objects are, and how it knows what it knows. The *ontological 'what'*, the pursuit of the objects studied by mathematics, and the *epistemological 'how'*, which concerns the way in which those objects should be studied, currently dominate the landscape of the philosophy of mathematics (Shapiro's square, figure 1, p.10). But curiously, the *'when'* and the *'why'*, which could allow mathematical philosophy to form stronger alliances with historical and phenomenological perspectives (chapters 10, 11), have mostly vanished into the horizon, at least within the Anglo-Saxon spectrum. This situation, however, must of necessity be a passing one, since there do not seem to be any intrinsic reasons for reducing the philosophy of mathematics to the philosophy of mathematical language. Everything rather points toward a far broader spectrum of *pendular practices*, *irreducible to imagination*, *reason or experience*, through which the conceptual evolution of the discipline will outstrip the sophisticated grammatical discussions that have been promoted by the analysis of language.

The traditional problems of the philosophy of mathematics have parceled themselves out around certain great dualities that have incessantly left their mark upon the development of philosophical reflection. Perhaps the unavoidable *crux* of the entire problematic lies in the deep feeling of wonder and astonishment that has always been a product of the 'unreasonable applicability' of mathematics to the real world. How can mathematics, this extraordinary human invention, grant us so precise a knowledge of the external world? The responses given have been numerous, carefully argued and, frequently, convincing. On the one hand, *ontological realism* has postulated that the objects studied by mathematics (whatever they are: ideas, forms, spaces, structures, etc.) lie buried in the real world, independently of our perception, while *ontological idealism* has suggested that mathematical objects are mere mental constructions. A realist stance thereby simplifies our supposed access to the real, while imposing strong constraints on the world (existential, formal, and structural constraints, etc.); in contrast, an idealist stance dismisses the world, sparing it from reliance on dubious organizational scaffoldings, but it faces the problem of mathematics' applicability head-on. *Epistemological realism*, on the other hand, has postulated (independently of any ontological position) that mathematical knowledge is not arbitrary and that its truth values are indices of a certain real stability, while *epistemological idealism* has regarded truth values as mere man-made mediations, which do not need to be propped up by any real correlate. An idealist stance again secures for itself a greater plasticity, with greater possibilities of access to the mathematical imagination, but it encounters serious difficulties at the junction of the imaginary and the real; a realist stance helps to understand mathematical thought's material success, but it places rigid restrictions on its creative liberty.

Interlaced with these basic, primary polarities, several other important and traditional dualities have found themselves in mathematical philosophy's spotlight. The *necessity* or *contingency* of mathematics, the *universality* or *particularity* of its objects and methods, the *unity* or *multiplicity* of mathematical thought, the *interiority* or *exteriority* of the discipline, the *naturalness* or *artificiality* of its constructions – each could count on having defenders and detractors of every sort. The status accorded to the correlations between physics and mathematics has always depended on the position one takes (whether consciously or unconsciously) with respect to the preceding alternatives. At the opposite extremes of the pendulum, we may situate, for example, a necessary, universal, unique and natural mathematics, very close to strongly realist positions, and a contingent, particular, multiple and artificial mathematics, coming very close, here, to the idealist extreme. But the *vast intermediate range* between these oscillations of the pendulum is ultimately what merits the most careful observation.[1] One of the principal objectives

---

1   An excellent overview of the entire pluralistic explosion of *philosophies* of mathematics can be found in G. Lolli, *Filosofia della matematica: L'eredità del novecento* (Bologna: il

of the present work is to demonstrate that – beyond a binary yes/no alternation – certain *mixtures* are *vital* for obtaining a thorough and accurate understanding of what it is to do mathematics, with respect to both its general and global structuration and many of its highly detailed, particular and local constructions.

In his excellent monograph, *Thinking about Mathematics*, Shapiro has made good use of a few of the aforementioned dualities in order to trace out a brilliant landscape of current philosophy of mathematics.[2] Restricting himself to the Anglo-Saxon world,[3] Shapiro goes on to classify several prominent bodies of work in virtue of their various realist or idealist stances (figure 1, based on Shapiro's text).[4]

Mulino, 2002). Lolli detects at least fourteen distinct currents (nominalism, realism, Platonism, the phenomenological tradition, naturalism, logicism, formalism, the semiotic tradition, constructivism, structuralism, deductivism, fallibilism, empiricism, schematism), in addition to a 'spontaneous philosophy' of mathematicians.

2   S. Shapiro, *Thinking about Mathematics: The Philosophy of Mathematics* (Oxford: Oxford University Press, 2000).

3   The restriction is not, however, explicit, and Shapiro commits the common Anglo-Saxon sin of believing that anything that has not been published in English does not form part of the landscape of knowledge. The identification of 'knowledge' with 'publication in English' has left outside of the philosophy of mathematics one who, to our understanding, is perhaps the *greatest* philosopher of 'real mathematics' in the twentieth century: Albert Lautman. For a discussion of 'real mathematics' (Hardy, Corfield) and the work of Lautman, see chapters 1–3.

4   Shapiro, *Thinking about Mathematics*, 32–3. Shapiro calls *realism in truth-value* the vision according to which 'mathematical statements have objective truth values, independent of the minds, languages, conventions, and so on of mathematicians' (ibid., 29). To simplify, we will here give the name 'epistemological realism' to this *realism in truth value*.

		EPISTEMOLOGY	
		Realism	*Idealism*
O N T O L O G Y	*Realism*	Maddy Resnick Shapiro	Tennant
	Idealism	Chihara Hellman	Dummet Field

Figure 1. *Contemporary tendencies in philosophy of mathematics, according to Shapiro.*

Aside from the differential details of the works included in the square above – a few of which we will compare with the results of our own investigations in part 3 of this essay – what we are interested in here is the *biparti-tion* that Shapiro elaborates. There is no place in this diagram for an ontological position *between* realism and idealism, nor for an epistemological *mixture* of the two polarities. Is this because such mediations are philosophi-cally inconsequential or inconsistent, or simply because they have been eliminated in order to 'better' map the landscape? One of our intentions in this essay will be to show that those mediations are not only consistent from a philosophical point of view (following Plato, Peirce and Lautman: chapter 3), but also *indispensable* from the point of view of contemporary mathematics. Shapiro's square therefore turns out to be only an ideal binary limit of a far more complicated, real state of affairs;[5] in an extended square, several new cells would appear, opening it onto *tertiary* frontiers.

Benacerraf's famous *dilemma* likewise presents itself by way of a dual alternative: *either* we coherently adopt a realism at once ontological and epistemological, and then find ourselves faced with difficult problems about how we might have knowledge of mathematical objects that do not originate in our own acts of invention and cannot be experientially perceived in nature; *or* we adopt a more flexible idealist epistemology and then find ourselves faced with other, equally difficult problems, as we inquire into the profound harmony between mathematics and the external world. But this either-or dilemma would not have to be considered as such if we could take stock of other intermediate positions between realism and idealism. We believe, in fact, that *mathematics* in its entirety produces illuminating examples of mediations between real and ideal configurations, and it does so from the most varied and complementary points of view (chapters 1, 4–7). Considered, as it usually is, from classical and dualistic perspectives, Benacerraf's Dilemma should be viewed with caution; however, considered within a broader metalogic, attentive to the dynamic evolution of mathematics – with

---

5   Just as, similarly, classical logic should in reality be understood as an *ideal limit* of intuitionistic logic. Cf. Caicedo's results, discussed in chapter 8.

its progressive osmoses and transferences between the real and the ideal – the dilemma falls apart, since there is no longer any reason to adopt dual exclusions of the either-or variety (chapters 8, 9).

It is important, here, to point out the dubious worth of fixing one's ontology and epistemology in advance, adopting them a priori before any observation of the mathematical universe, and presuming to impose certain rigid partitions upon the latter. At the very least, such an adoption of philosophical presuppositions prior to even setting eyes on the mathematical world has limited our perspective and has led to the perception of a rigid, static and eternal mathematics, a perception that has little or nothing to do with the *real* mathematics that is being done every day. Instead, a living mathematics, in incessant evolution, should be considered the basic *presupposition* of any *subsequent* philosophical consideration whatsoever. The study of the continuities, obstructions, transfers and invariants involved in *doing mathematics* should then – and only then – become an object of philosophical reflection. The elaboration of a *transitory* ontology and epistemology, better matched to the incessant *transit* of mathematics, is the order of the day.[6] The peerless strength of mathematics lies precisely in its exceptional *protean* capacity, a

---

6   Alain Badiou explores this idea in his *Court traité d'ontologie transitoire* (Paris: Seuil, 1998). [Tr. N. Madarasz as *Briefings on Existence: A Short Treatise on Transitory Ontology* (Albany, NY: SUNY Press, 2006).] Part 3 of the present study makes further inroads into that 'transitory philosophy' that, we believe, mathematics demands.

remarkable transformative richness that has rarely been philosophically assimilated.

One of the traditional problems that mathematical philosophy has had to confront, in this respect, has to do with the general place of mathematics within culture as a whole. Here, a dualistic reading is again burdened with immediate problems: If mathematics is understood as an evolutive forge, internal to *contingent* human creativity,[7] the problem arises as to how we can explain its apparently *necessary* character and its cumulative stability; if, instead, mathematics is understood as the study of certain forms and schemas that are independent of its cultural environment,[8] there arises the problem as to how we can explain the markedly historical character of mathematical 'discoveries'. In practice, a *middle way* between both options seems far better adjusted to the reality of doing mathematics (see, in particular, the meditations on Grothendieck in chapter 4): a fluctuating, evolving activity, full of new *possibilities*, springing from disparate cultural realms, but always managing to construct precise *invariants* for reason behind the many relative obstructions that the mathematical imagination is always encountering. What drives both mathematical creativity and its subsequent normalization is a to and fro that tightly

---

7   This is the case, for example, in R. L. Wilder, *Mathematics as a Cultural System* (Oxford: Pergamon Press, 1981).

8   M. Resnik, *Mathematics as a Science of Patterns* (Oxford: Oxford University Press, 1997).

weaves together certain sites of pure possibility and certain necessary invariants, within well-defined contexts.

Without this *back-and-forth* between obstructions and invariants, mathematics cannot be understood. The wish to reduce, a priori, the doing of mathematics to one side of the balance or the other is, perhaps, one of the major, basic errors committed by certain philosophers of mathematics. The *transit* between the possible, the actual and the necessary is a strength *specific* to mathematics, and one that cannot be neglected. To consider that transit as a weakness, and to therefore try to eliminate it, by reducing it either to contingent or to necessary circumstances (another version of an either-or exclusion), is an unfortunate consequence of having taken sides in advance, *before* observing the complex modal universe of mathematics. In fact, as we shall demonstrate in part 2 of this book, on the basis of the case studies of part 2, in mathematics, discovery (of necessary structural schemas) is *just as* indispensable as invention (of languages and possible models). The tight mathematical weave between the real and the ideal *cannot be reduced* to just one of its polarities, and it therefore deserves to be observed through a conjunction of complementary philosophical points of view. We believe that any reduction at all, or any preemptive taking of sides, simply impedes the contemplation of the specificities of mathematical transit.

Both Wilder and Resnik, to point to just one complementary polarity, have much to offer us. A hypothesis (chapter 1), a program (chapter 3) and a few detailed case studies (chapters 4–7) will prepare us with some outlines for a synthesis (chapters 8–11) by which various central aspects of complementary perspectives, like those of Wilder and Resnik, can come to be 'glued' together in a unitary whole. We may point out that one of the essential and basic motivations of this work is the desire to elaborate, in order to reflect on mathematics, a sort of *sheaf* that would allow us to reintegrate and 'glue together' certain complementary philosophical viewpoints. As will become clear in part 2, the notion of a *mathematical sheaf* is probably the fundamental distinguishing concept around which the elaboration of contemporary mathematics, with new impetus, begins, with all of its extraordinary instruments of structuration, geometrization, gluing, transfer and universalization – and so the attempt to look at mathematics *from a sheaf of equally complex perspectives* turns out to be a rather *natural* one. To achieve this, we will have to delimit certain 'coherence conditions' between complementary philosophical perspectives (chapters 1, 3) in order to then proceed with a few sketches of 'sheaving' or of 'structural synthesis' (chapters 8–11).

ACKNOWLEDGEMENTS

To Pierre Cassou-Noguès, Marco Panza and José Ferreirós, who, with their invitation to Lille in 2005, helped me to confirm continental philosophy's crucial role in achieving a greater comprehension of advanced mathematics, and also to define the singular interest that this monograph might gather in a field rather flattened by analytic philosophy. To Carlos Cardona, who, with his brilliant doctoral thesis on Wittgenstein and Gödel, provoked me to contradict him, thereby giving rise to the solid kernels of this work. To my students in Lógica IV (2006), who, during this text's gestation, stoically bore the brunt of some of its most abstract assaults. To my colleges and participants in my Seminario de Filosofía Matemática, whose constructive criticisms gave rise to various lines of thought inscribed in this text. To Juan José Botero, who, with his invitation to present as a plenary speaker in El Primer Congreso Colombiano de Filosofía (2006), opened an uncommon space to me, one that other colleagues (mathematicians, philosophers, cultural scholars) carefully ignored. To Andrés Villaveces, who proposed some remarkable clarifications in the course of writing this text, and who unconditionally supported me with his always-magnificent enthusiasm. To Xavier Caicedo, who produced a profound and encouraging review of the monograph, supporting my (unsuccessful) candidacy for El Premio Ensayo Científico Esteban de Terreros (2008). To Javier de Lorenzo, whose friendship and generosity came very close to placing this text with an improbable commercial publishing house in his country. To Alexander Cruz, Magda González, Epifanio Lozano, Alejandro Martín and Arnold Oostra, who corrected many of the work's glitches and improved several of its paragraphs. To La Editorial Universidad Nacional, who granted me the opportunity of entering, by contest, its fine collection, Obra Selecta.

My greatest thanks go to Zachary Luke Fraser, who not only worked carefully on a daunting translation, but in many paragraphs improved on the original. Finally, heartfelt thanks go to my editors at Urbanomic, Robin Mackay and Reza Negarestani, whose great vision is opening a wide range of much-needed doors for new forms of contemporary thought.

PART ONE

The General Environment of Contemporary Mathematics

CHAPTER 1

THE SPECIFICITY OF MODERN AND CONTEMPORARY MATHEMATICS

It is well known that mathematics is presently enjoying something of a boom. Even conservative estimates suggest that the discipline has produced *many* more theorems in the last three decades than it has in its entire preceding history, a history stretching back more than two thousand years (including the very fruitful nineteenth and twentieth centuries, and all the way up to the 1970s). The great innovative concepts of modern mathematics – which we owe to Galois, Riemann and Hilbert, to cite only the three major foundational figures – have been multiplied and enriched thanks to the contributions of a veritable pleiad of exceptional mathematicians over the last fifty years. Proofs of apparently unattainable theorems – like Fermat's Last Theorem, or the Poincaré conjecture – have been obtained, to the surprise of the mathematical community itself, thanks to the unrelenting struggle of mathematicians who knew how to carefully harness the profound explorations already undertaken by their colleagues. The boom in mathematical publications and reviews appears unstoppable, with an entire, flourishing academic 'industry' behind it; though the excessive number of publications could be taken to discredit their quality (it is tempting to say that publication should be cause for penalty rather than promotion), the immense liveliness of mathematics is made manifest in the frenetic activity of publishing houses. Meanwhile, mathematics' relations with physics again seem to have found a moment of grace, as the two are profoundly interlaced in the study of superstrings, quantizations, and complex cosmological models.

Strangely, however, the *philosophy of mathematics* has rarely taken stock of the genuine *explosion* that mathematics has witnessed in the last fifty years (chapter 2). Two reasons may be given for this: firstly, the view that, in spite of the advance and evolution of mathematics itself, mathematics' methods and the *types* of objects it studies remain invariable; secondly, the simple myopia before the new techniques and results, because of a certain professional incapacity to observe the new thematics at stake. In practice, in fact, there seems to be a *mutual feedback* between these two tendencies; on the one hand, the conviction that set theory (and with variants of first-order logic) already provide sufficient material for the philosophy of mathematics to shore up an unwillingness to explore other environments of mathematical knowledge; on the other hand, the inherent difficulty involved in the advances of *modern mathematics* (from the second half of the nineteenth century through the first half of the twentieth), and, a fortiori, those of *contemporary mathematics* (from the middle of the twentieth century onwards), can be avoided by hiding behind the supposed ontological and epistemological invariability of the discipline. This deliberate neglect of the discipline's (technical, thematic, creative) landscape is a situation that would be seen as *scandalous* in the philosophy of other scientific disciplines,[9] but to which the philosophy of mathematics seems able to resign itself, with a restrictive security and without the least bit of modesty.

Two great extrapolations – equivocal by our lights, as we shall attempt to demonstrate throughout this essay – support the idea, ubiquitous in the philosophy of mathematics, according to which it is *unnecessary* to observe the current advances of the discipline. On the one hand, the objects and methods of *elementary mathematics* and *advanced mathematics* are considered not to essentially differ from one another; on the other hand, the development of mathematics is presupposed to have a markedly necessary character and an absolute background. If, from an epistemological and ontological point of view, the exploration of the Pythagorean Theorem offers nothing

---

9   A philosophy of physics that does not take stock of the *technical* advances in physics, for example, would be unthinkable. B. d'Espagnat, *Le réel voilé. Analyse des concepts quantiques* (Paris: Fayard, 1994), for instance, performs an admirable philosophical study of quantum physics, in which the notable technical advances of the discipline are carefully observed, and in which it is demonstrated that, in order to understand quantum physics, *new* ontological and epistemological approaches, *adapted* to the new methods and objects of knowledge, are required.

different than the exploration of Fermat's, then making the effort to (philosophically) understand all the instruments of algebraic and complex variable geometry that opened a way to proving Fermat's Theorem would, of course, be pointless. If, from a historical and metaphysical point of view, the evolution of mathematics is considered not to give rise to new types of 'entities', then it would be equally absurd to try to entangle oneself in the complexities of contemporary mathematical creativity. We nevertheless believe that these two ubiquitous suppositions – that there is no distinction between elementary and advanced mathematics; that there is no duality of transits and invariants in mathematics – are only valid, *in part*, in determinate, restrictive contexts, and we consider the extrapolations of these suppositions into the 'real' totality of mathematics (and contemporary mathematics, in particular), to constitute a profound methodological error.

Following David Corfield, we will call 'real mathematics' the warp of advanced mathematical knowledge that mathematicians encounter daily in their *work*,[10] a warp that can be seen as perfectly *real* from several points of view: as a stable object of investigation for a broad community, as an assemblage of knowledge with a visible influence

---

10 D. Corfield, *Towards a Philosophy of Real Mathematics* (Cambridge: Cambridge University Press, 2003). As Corfield has pointed out, Hardy, in his polemical *A Mathematician's Apology*, called 'real mathematics' the mathematics constructed by figures like Fermat, Euler, Gauss, Abel and Riemann (G.H. Hardy, *A Mathematician's Apology* [Cambridge: Cambridge University Press, 1940], 59–60; cited in Corfield, 2).

on the practice of the discipline, and as a framework that can be effectively contrasted with the physical world. Both elementary mathematics and set theory, the object of extensive considerations in analytic philosophy, are thus but slight fragments of 'real' mathematics. They considerably extend the scope of *classical mathematics* (midseventeenth to midnineteenth century), but in view of the totality (figure 2, overleaf), it must be observed that, nowadays, modern and contemporary mathematics largely make up the core of the discipline. Let us take as a *basic supposition* the insufficiently appreciated fact that apprehending the *totality* of mathematical production with greater fidelity and technical precision may turn out to be of great relevance for philosophy.

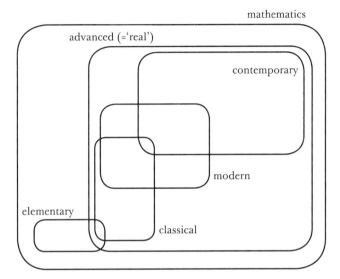

Figure 2. *Correlations between the areas of mathematics: elementary, advanced, classical, modern, contemporary.*

The boundaries that allow us to distinguish the aforementioned areas are clearly historical, since leading mathematical research *becomes progressively more complex* throughout its evolution. Nevertheless, the boundaries can also be associated with certain types of mathematical instruments, introduced by great mathematicians, whose names still serve to characterize each epoch:

Classical mathematics (midseventeenth to midnineteenth centuries): sophisticated use of the infinite (Pascal, Leibniz, Euler, Gauss);

Modern mathematics (midnineteenth to midtwentieth centuries): sophisticated use of structural and qualitative properties (Galois, Riemann, Hilbert);

Contemporary mathematics (midtwentieth century to present): sophisticated use of the properties of transference, reflection and gluing (Grothendieck, Serre, Shelah).

In particular, there accumulated in modern mathematics an enormous quantity of knowledge, which evolved and went on to make up to the current body of mathematics: set theory and mathematical logic, analytic and algebraic number theory, abstract algebras, algebraic geometry, functions of complex variables, measure and integration, general and algebraic topology, functional analysis, differential varieties, qualitative theory of differential equations, etc.[11] Even if a series of important mathematical theorems have succeeded in proving that *any* mathematical construction can be represented inside a suitable set theory

---

11 The Mathematical Subject Classification 2000 (*MSC 2000*) includes some *sixty* principal entries in a tree that goes on to rapidly branch out. Above, we indicate only a few of the initial *indispensable* entries in the tree.

(and that the enormous majority of mathematics can be represented inside the Zermelo-Fraenkel theory, with its underlying first-order classical logic), it is nevertheless clear, within mathematical practice, that what value these 'facsimiles' have is merely logical, and *far removed* from their genuine *mathematical value*. We believe the fact that mathematical constructions can be reduced, theoretically, to set-theoretical constructions has been enlisted as yet another prop by means of which, in the philosophy of mathematics, one could, for so long, avoid a more engaged inspection of 'real mathematics'. Nevertheless, as we shall soon see, the *structures* at stake and the ways of doing things differ dramatically between set theory and other mathematical environments, and the ontology and epistemology we propose should, consequently, differ as well (to say nothing of history or 'metaphysics' – chapters 10–11). The *possibility* of reducing the demonstration of a complex mathematical theorem to a series of purely set-theoretical statements (a *possibility* existing only in theory and never executed in practice, as soon as certain rather basic thresholds are crossed) has, in the philosophy of mathematics, been elevated to a fallacious extrapolation, an extrapolation that has allowed certain philosophical perspectives to shirk any investigation into the *present* of 'real mathematics', beyond mathematical logic or set theory.

The environment of advanced mathematics, already clearly delimited by the middle of the twentieth century, found an exceptional philosopher in Albert Lautman.[12] For Lautman, mathematics – beyond its *ideal* set-theoretical reconstruction – hierarchizes itself into *real* environments of dramatically varying complexity, where concepts and examples are interlaced through processes that bring the *free* and the *saturated* into structural counterpoint with one another, and where many of the greatest mathematical creations emerge through the mediation of *mixtures*. Entering into the vast conglomerate of the mathematics of his time, Lautman was able to detect certain features *specific* to advanced mathematics,[13] features that *do not appear* in elementary mathematics:

---

12 The work of Albert Lautman (1908–1944) deserves to be understood as the most incisive philosophical work of the twentieth century that both situated itself within *modern mathematics* and sought to outline the hidden mechanisms of advanced mathematical creativity, while synthesizing the *structural and unitary* interlacings of mathematical knowledge. Lautman's writings, forgotten and little understood at present, have resurfaced in a new French edition (A. Lautman, *Les mathématiques, les Idées et le Réel physique*, [Paris: Vrin, 2006]), a recent English translation (tr. S. Duffy as *Mathematics, Ideas and the Physical Real* [London: Continuum, 2011]) and in the first complete translation of his works into another language (tr. F. Zalamea as *Ensayos sobre la dialéctica, estructura y unidad de las matemáticas modernas* [Bogotá: Universidad Nacional de Colombia, 2011]). For a critical presentation of Lautman's work, see my extensive scholarly introduction to the Spanish edition. In the present essay, I aim to develop Lautman's work somewhat, and extend its scope from modern mathematics (as known to Lautman) to contemporary mathematics (which now lie before us).

13 The critical works attending to the multiplicity of advanced mathematical creations are few in number, and so it's worth calling attention to a work so kindred to Lautman's as that of Javier de Lorenzo, who has always been attentive to the deep strata and diverse ramifications of modern mathematical invention. Among his works, see, in particular, *Introducción al estillo mathemático* (Madrid: Tecnos, 1971); *La matemática y el problema de su historia* (Madrid: Tecnos, 1977); *El método axiomático y sus creencias* (Madrid: Tecnos, 1980); *Filosofías de la matemática fin de siglo XX* (Valladolid: Universidad de Valladolid, 2000). Lorenzo does not seem to be familiar with Lautman, nor does he mention him in his writings.

1. *a complex hierarchization* of diverse mathematical theories, irreducible to one another, relative to intermediary systems of deduction;

2. *a richness* of models, irreducible to merely linguistic manipulations;

3. *a unity* of structural methods and conceptual polarities, behind their effective multiplicity;

4. *a dynamics* of mathematical activity, contrasted between the free and the saturated, attentive to division and dialectics;

5. *a theorematic interlacing* of what is multiple on one level with what is one on another, by means of mixtures, ascents and descents.

We should contrast elementary mathematics – the privileged focus of analytic philosophy – with the advanced mathematical theories that make up the sweeping spectrum of modern mathematics. An often-repeated argument for the possibility of reducing the scope of inquiry to that of elementary mathematics comes down to assuring us that every mathematical proposition, since it is a tautology, is equivalent to every other, and so, from a philosophical perspective, it is enough to study the spectrum of elementary propositions. For example, the extremely anodyne '2+2=4' would, from a logical point of view, be equivalent to the significant and revealing Hahn-Banach Theorem (HB), since both propositions are deducible from the Zermelo-Fraenkel system of axioms (ZF). Nevertheless, the 'trivial' tautological equivalence $ZF \vdash HB \leftrightarrow 2+2=4$ is as far from exhausting the mathematical content of the theorems as it is from exhausting their logical status. The equivalence effectively collapses as soon as, instead of starting with ZF, we opt for *intermediate* axiomatic systems. In fact, Friedman and Simpson's *reverse mathematics*[14] show that the basic propositions of arithmetic (which, for example, Wittgenstein repeatedly studies in his *Lectures on the Foundations of Mathematics*)[15] show up on the lowest levels of mathematical development (in the system RCA_0, reduced to demonstrating the existence of recursive sets), while the HB not only requires more advanced instruments (a WKL_0 system with weak forms of König's lemma), but is fully *equivalent* to those instruments. To be precise, it turns out that $RCA_0 \nvdash HB \leftrightarrow 2+2=4$, since we have $RCA_0 \vdash HB \leftrightarrow WKL_0$, $RCA_0 \vdash 2+2=4$ and $RCA_0 \nvdash WKL_0$.

The consequences of this state of affairs are obvious, but they have not been sufficiently considered in the philosophy of mathematics. First of all, it seems absurd to

---

14 S. G. Simpson, *Subsystems of Second Order Arithmetic* (New York: Springer, 1999).

15 L. Wittgenstein, *Lectures on the Foundations of Mathematics, Cambridge 1939* (Chicago: University of Chicago Press, 1989).

compare pairs of mathematical propositions *with respect to* excessively powerful base systems. *In the eyes of ZF*, all demonstrable propositions are logically trivialized (as pairs of equivalent tautologies) – not because the propositions contain an identical logical (or mathematical) value in themselves, but because the differences are not appreciated *by* ZF. ZF is a sort of deductive *absolute*, in which both those who study the set-theoretical universe and those who wish to restrict themselves to elementary mathematics alone can be quite comfortable; nevertheless, the intermediate *thresholds* of deductive power in ZF (like the systems studied in *reverse mathematics*) constitute the genuinely relevant environments from the point of view of 'real' mathematics, with the multiple *hierarchies* and differences in which one may carry out a mathematically productive study of logical obstructions and transfers. Secondly, the idea of a tautological mathematics, fully expressible within the narrow scope of elementary mathematics, seems untenable. The moment we cross the complexity thresholds of system RCA_0 (and pass over into system ACA_0, in which we can prove the first important results of abstract algebra, like the existence of maximal ideals in commutative rings), we enter into a *relative web* of partial equiconsistencies where the (supposedly stable and absolute) notion of tautology is deprived of any real mathematical sense. Mathematics goes on producing necessary theorems, but within variable deductive contexts, whose oscillations and changes are fundamental to the expression of the theorems' true mathematical value. Thirdly, the vital presence, in the discipline, of certain logical and mathematical *irreducibilities* becomes palpable. Mathematics' richness takes root in its *weave* of demonstrations (the impossibility of evading certain obstructions and the possibility of effecting certain transfers), something that unfortunately disappears in the light of *extreme* perspectives – whether from an absolutely tautological perspective (ZF, where all is transferable) or from elementary perspectives (subsystems of RCA_0, where all is obstruction).

The *complex hierarchization* of advanced mathematics (point 1, noted above) gives rise to a panoply of constructive scales, inverse correspondences and gradations of every kind (particularly visible in Galois theory and in the generalized theories of duality), which allow the emergence of mathematical creativity to be studied with greater *fidelity*. The blooming and genesis of mathematical structures, hidden from a static, analytic approach, are better seen from a dynamic perspective, in light of which a problem, concept or construction *is transformed* by the problem's partial solutions, the concept's refined definitions, or the construction's sheaf of saturations and decantations.[16] In that eminently living and incessantly

---

16 The images of decantation, transfusion and distillation that recur throughout this work indicate those creative gestures by way of which, as we shall see, mathematical ideas or structures are 'poured' – sometimes with the help of others, as 'filters' – from one register to another, often leaving behind, as a kind of 'sediment', features previously thought to be integral to them.

evolving field of thought that is mathematics, a profound hierarchization is not only indispensable to, but is the very *engine* of creation. Any philosophy of mathematics that fails to take stock of the complex, hierarchical richness of advanced mathematics will be led to neglect not only such delicate 'intermittencies of reason' (differences in logical contrast), but the even subtler 'intermittencies of the heart' (differences in mathematical creativity).

Among the features that distinguish modern from elementary mathematics, Point 2 likewise harbors robust philosophical potential. Modern mathematics has produced, in all of its fields of action, really remarkable conglomerates of models, extremely diverse and original, with significant structural distinctions. It deals, in fact, with *semantic* collections that greatly surpass the more restrained syntactic theories that these collections help shape, as can be seen, for example, when we compare the explosive and uneven universe of simple finite groups with the elementary axiomatization that underlies the theory. Advanced mathematics contains a great semantic richness, irreducible to merely grammatical considerations, though a *fallacious extrapolation* has presumed to identify the making of mathematics with the making of certain grammatical rules. Indeed, the supposed reduction of mathematical thought to a deductive grammar is understandable from the point of view of elementary mathematics, where the models tend to be few and controlled, but it is a monumental trivialization to extrapolate that situation to 'real' mathematics, where classes of models start to behave in an altogether erratic fashion (see chapter 5, on Shelah's works). The attempt to reduce mathematics to grammar, in short, assumes a (fallacious) reduction of mathematics to elementary mathematics and then applies the (plausible) identification of elementary mathematics with finitary grammatical rules.

The richness of modern mathematics is, to a large extent, rooted in the enormous diversity of structures and models that have been constructed (*or* discovered – we won't get into the question for now, though we believe that *both* construction *and* discovery are indispensable; see c,hapters 8, 9). Structures of every sort have indelibly furrowed the current landscape of mathematics, and a clear and distinctive mark of advanced mathematics consists in having to *simultaneously* consider multiple structures in any comprehensive inspection of a mathematical phenomenon. The phenomenon frequently demands to be considered under complementary points of view, whereby quite diverse arithmetical, algebraic, topological and geometrical instruments crisscross each other. A fundamental characteristic of modern mathematics is its capacity to operate *transfusions* between a multiplicity of apparently discordant structures, exploiting remarkable sets of instruments that succeed in harmonizing the diversity.[17] Without variety, multiplicity and complexity, modern mathematics would not have even been able to emerge; and, as we shall see, without interlacing and unity, it would not have been able to consolidate itself. The situation is very different than that of elementary mathematics, where structures are strictly determined – the integers, the real plane, and little else – and, for that reason, are unable to give rise to either a variability of models or a fluxion between mathematical subdisciplines. This, again, is a key observation: Though a restriction to elementary mathematics might allow for a conflation of models and language, and the elimination of semantics' variability and fluxion, this sort of thing must be abandoned as soon as we enter into advanced mathematics, where the landscape is firmly governed by collections of (mathematical) structures and (physical) facts, often independently of any syntactic or linguistic considerations.

The multiplicative and differential richness of modern mathematics is accompanied by a complementary, pendular tendency toward the unitary and the integral (point 3, indicated above). The dialectical tensions between the One and the Multiple have found, in modern mathematics, a fertile field of *experimentation*. The unity of mathematics expresses itself, not only in virtue of a common base upon

---

17 Thus responding to the first epigraph by Goethe that appears at the beginning of this study.

which the All is reconstituted (set theory), but – before all else – in the convergence of its methods and in the *transfusing* of ideas from one to another of its various webs. The penetration of algebraic methods into analysis, itself subordinated to topology, the ubiquitous geometrization of logic and the structural harmony of complex analysis with arithmetic, are all examples in which mathematics' global unity can be perceived in its local details. A profound epistemological inversion shows how – contrary to what we might think at first – an attentive observation of practical diversity *permits* a later reintegration of the One behind the Multiple. In fact, a full awareness of diversity does not reduce to the disconnected, but rather turns back to unity, whether in Peirce's pragmatism, Benjamin's montage, Francastel's relay or Deleuze's difference. Similarly – and with great technical precision, as we shall see in chapters 4 and 7 – modern mathematics seeks (and finds) ways to concatenate a prolific multiplicity of *levels* into great *towers* and unitary frameworks.

That *reconstruction* of the One behind the Multiple is another of the fundamental marks allowing us to separate elementary mathematics from advanced mathematics. Elementary mathematics is 'one' from the outset, because it has not yet managed to multiply or differentiate itself; advanced mathematics, by contrast, having already passed through explosively creative processes, has had to relearn and reconstruct common ties and warps in the midst of diversity. The firmness and solidity produced by that double weaving movement – differentiation/integration, multiplication/unification – are virtues *proper* to advanced mathematics, which can be only very faintly detected in elementary mathematics. In fact, some of the great unitary theories of contemporary mathematics – generalized Galois theory, algebraic topology, category theory – are *trivialized* on the elementary level, since the structures at play fail to achieve enough differential richness to merit any subsequent reintegration. It is factually impossible, therefore, to claim to observe the same kinds of conceptual movements in reasoning over tally marks as one finds when, for example, one enters into the theory of class fields. We believe that failures to understand or assume this sort of distinction have done quite enough damage to the philosophy of mathematics.

Immediately bound up in the weave between multiplicity and unity in modern mathematics, we find the inescapable dynamism of doing mathematics (point 4). Mathematics, developing as it has from the middle of the nineteenth century up to the present day, has not ceased to create new spaces for the understanding. A pendular process – in which, on the one hand, meticulous saturations within *particular* structures accumulate, and on the other, the free behavior of *generic* structures is set loose – allows for the simultaneous contemplation of an uncommonly precise spectrum of local obstructions/resolutions and a series of global organizational schemas. The dynamic transit between the local and the global is one of the major successes of modern mathematics, a transit that is hard to perceive in elementary mathematics, in which a clear preponderance of the local takes precedence. Again, there seems to be an unwarranted extrapolation at work when one presumes to take the eminently static, finished, stable and 'smooth' character of elementary mathematics as characteristic of *all* of mathematics in its entirety. Advanced mathematics are, by contrast, essentially dynamic, open, unstable, 'chaotic'. It is not by chance that, when one asks mathematicians about the future of their discipline, almost all of them leave the landscape completely open; with a thousand forces pulling in different directions, the 'geometry' of mathematical creativity is replete with unpredictable singularities and vortices.

The back-and-forth between diverse perspectives (conceptual, hypothetical, deductive, experimental), diverse environments (arithmetical, algebraic, topological, geometrical, etc.) and diverse levels of stratification within each environment is one of the fundamental dynamic features of modern mathematics. When that pendular back-and-forth partially concretizes itself in theorematic warps and interlacings, and when the transit of ascents and descents between certain levels of stratifications – along with a great arsenal of intermediate *mixtures* to *guide* the transit – is systematized, we then find ourselves faced with (point 4) other peculiarities, specific to advanced mathematics. In fact, at low levels of complexity, such as those found in elementary mathematics, the elevations (ascents/descents) and intermediate constructions (mixtures) *naturally* tend to trivialize away and vanish. It was necessary, for example, for obstructions in infinitary systems of linear equations and in classes of integral equations to be confronted in order for the notion of a Hilbert space, one of modern mathematics' most incisive mixtures, to emerge, just as certain singularities in complex variable functions had to be confronted for another paradigmatically modern construction, the notion of Riemann surfaces, to emerge. Similarly, Galois theory – one of the great buttresses of mathematics' development, with remarkable conceptual transfers into the most varied mathematical domains – would be unthinkable had important obstructions between webs of notions associated with algebraic solutions and geometrical invariants not been taken into account. In order to tackle problematics of great complexity – stretched over highly ramified dialectical warps – modern mathematics finds itself obliged to *combine* multiple mathematical perspectives, instruments and bodies of knowledge, something that rarely happens in the realms of elementary mathematics.

Beyond points 1–5, which we have just discussed,[18] and which constitute an initial plane of separation between elementary and modern mathematics (from the middle of the nineteenth century to the middle of the twentieth, as we have defined it), we believe that *contemporary* mathematics (1950 to the present) incorporates additional criteria that reinforce its specificity. *Beyond conserving* those distinctively modern characteristics (1–5),[19] contemporary mathematics bears new, distinctive elements, as compared to elementary mathematics, among which we may point out the following:

6. the structural *impurity* of arithmetic (Weil's conjectures, Langlands's program, the theorems of Deligne, Faltings and Wiles, etc.);

7. the systematic *geometrization* of all environments of mathematics (sheaves, homologies, cobordisms, geometrical logic, etc.);

18 Lautman's work (see note 12 and chapter 2) provides a great variety of technical examples, concretizing the aforementioned tendencies, as well as other formulations of points 1–5.

19 There is no 'postmodern break' in mathematics. Following Rodríguez Magda, it is far more appropriate to speak of transmodernity than of a dubious 'post'-modernity, when seeking to characterize our age. (R. M. Rodríguez Magda, *Transmodernidad* [Barcelona: Anthropos, 2004]). In mathematics – and, in fact, in culture as a whole, as we have remarked in our essay *Razón de la frontera y fronteras de la razón* (F. Zalamea, *Razón de la frontera y fronteras de la razón: pensamiento de los límites en Peirce, Florenski, Marey, y limitantes de la expresión en Lispector, Vieira da Silva, Tarkovski* [Bogotá: Universidad Nacional de Colombia, 2010]) – continuous notions connected with traffic and frontier are indispensable. The prefix 'trans-' seems, therefore, far more indicative of our condition (and of the mathematical condition) than a premature 'post-'.

8. the *schematization*, and the liberation from set-theoretical, algebraic and topological restrictions (groupoids, categories, schemas, topoi, motifs, etc.);

9. the *fluxion* and deformation of the usual boundaries of mathematical structures (nonlinearity, noncommutativity, nonelementarity, quantization, etc.);

10. the *reflexivity* of theories and models onto themselves (classification theory, fixed-point theorems, monstrous models, elementary/nonelementary classes, etc.).

Many of the major innovative works of the great contemporary mathematicians[20] can be situated, *grosso modo*, along the aforementioned lines, as we suggest in the following table:[21]

---

20 The selection is, inevitably, personal, though the list indubitably includes some of the fundamental figures of mathematics since 1950. We only include in the table those mathematicians whom we are studying in the second part of this essay. (The order of appearance in the table corresponds to the order in which each author is studied in the second part of our essay.) Other indispensable figures of contemporary mathematics do not appear here (such as Borel, Chevalley, Dieudonné, Drinfeld, Eilenberg, Gelfand, Margulis, Milnor, Smale, Thom, Thurston and Weil, to name just a few), since, in most cases, we mention them only in passing, without dedicating a specific section to their works.

21 The marks indicate a *clear preponderance* of works along each line, and not mere incursions that might be considered limited in comparison with the remainder of the work of the mathematician in question. Grothendieck is clearly situated above all other mathematicians of the last half century, as is faintly indicated by the *five* marks that serve to register the enormous presence of his work. The other marks should be understood as merely indicative, though they are also adequately representative.

	6	7	8	9	10
Grothendieck	•	•	•	•	•
Serre	•	•	•		
Langlands	•	•	•		
Lawvere		•	•	•	
Shelah		•	•	•	•
Atiyah		•	•	•	
Lax		•		•	
Connes	•	•	•	•	
Kontsevich		•	•	•	
Freyd		•	•		•
Simpson			•		•
Gromov		•	•	•	
Zilber		•	•	•	

Figure 3. *A few great mathematicians and their contributions to the major lines of development of contemporary mathematics.*

Behind arithmetical mixing (6), geometrization (7), schematization (8), structural fluxion (9) and reflexivity (10), we find some modes of conceptualization and construction pertaining to contemporary mathematics that are not in evidence (or that appear only *in nuce*) in the period from 1900–1950. An initial, fundamental inversion consists in studying fragments of mathematics, by setting out not from partial axiomatic descriptions (as in Hilbert's program), but from classes of correlated structures.

For both mathematical logic (with the unprecedented blossoming of model theory) and pure mathematics (with category theory), the objects studied by mathematics are not only collections of axioms and their associated models, but also, from an *inverse perspective*, classes of structures and their associated logics (a point of view that is indispensable for the *emergence* of abstract model theory and generalized quantifiers, after Lindström). In cases where the class of structures is very extensive and runs transversally through many fields of mathematics (such as the intermediate categories between regular categories and topoi), the breadth of perspective often provides for new global theorems (synthetic hierarchization, delimitation of frontiers, transference – as in Freyd's representation theorems). In cases where the class arises on the basis of certain particular structures and their infinitesimal deformations (as in quantization), a precise and profound apprehension of the class brings with it local technical advances of a remarkable nature (analytic decomposition, fluxion, asymptotic control – as in Perelman's proof of the Poincaré conjecture). In either case, however, mathematics explicitly *comes back to precede* logic. The situation we are dealing with here is a basic one (broadly prefigured by Peirce, to whom we shall return), which may, in fact, have always subsisted in mathematical *practice*, but which was, again and again, hidden from the perspectives available to analytic philosophy or the philosophy of language in the twentieth century.

A second essential inversion has to do with contemporary mathematics' tremendous capacity to construct incisive technical breaches of apparently insuperable boundaries – nonelementary classes (Shelah), noncommutative geometry (Connes), nonidempotent logic (Girard's nonperennial, or *linear*, logic), etc. – going beyond the normalized environments that had arisen naturally in the discipline during the first half of the twentieth century. Instead of progressing from a positive interior, in which knowledge is accumulated, toward a negative, somehow unknowable exterior, contemporary mathematics sets itself within the determinate *boundaries* of the '*non-*' from the outset, and, setting out from those frontiers, goes on to constructively explore new and astounding territories. A third inversion consists in considering mathematical *mixtures*, not as intermediary entities that are useful in a deduction, but as proper, original entities, in which the very construction of the discipline is at stake. There is no space of contemporary mathematics that does not find itself mongrelized by the most diverse techniques; the mediatory ('*trans-*') condition, which in the first half of the twentieth century could be seen as one step in the path of demonstration, is today becoming the very core of the discipline. The extraordinary *combination* of arithmetical structuration, algebraic geometrization, schematization, fluxion and reflexivity in Grothendieck's work is a prime example, in which every instrument is simply directed toward controlling the *transit* of certain global mathematical conceptions through an enormous spectrum of local environments.

We should note that, in every case, the oscillations and inversion remarked upon do not appear in elementary mathematics, and indeed cannot appear in the latter's restricted fields of action. Moreover, before a sweeping set-theoretical landscape such as ZF with first-order classical logic, many of the aforementioned currents become 'non-observables' of some sort. One of the basic deficiencies of mathematical philosophy has consisted in *not coupling* its philosophical instruments of observation with the environments observed, and in attempting to paint standardized landscapes of the whole. Another methodology (one more in tune with the development of contemporary mathematics) could consist in observing certain environments of mathematics through philosophical filters more *adequate* to them – and *then* trying to synthetically *glue* the diverse philosophical observations thus obtained. (See chapter 3 for the pragmatic – in Peirce's sense – program that can be sketched out in this way, and chapters 8–11 for the realization of certain partial gluings.)

If it is entirely *natural* to use the instruments of analytic philosophy in order to view the set-theoretical universe, with its underlying first-order classical logic, it seems rather wrongheaded to carry on with the same old instruments in the study of *other* mathematical environments (other environments that constitute the *majority*: set theory occupies only a very limited space of mathematical investigation, as can be seen in the *MSC 2000*). As we shall see later on, certain *dialectical* instruments are indispensable for capturing schematization and fluxion, just as it is only from a synthetic and *relational* perspective that the great currents of contemporary structuration can be understood, and only from a fully *modal* perspective that the inexhaustible richness of the continuum can be observed. The specificity of modern and contemporary mathematics *forces* us to keep changing our filters of philosophical observation, and so we find the elaboration of a new *philosophical optics* coming into play, one that – with a pragmatic machinery of lenses to insure a minimum of distortions – will allow us to survey the landscape of 'real' mathematics.

CHAPTER 2

ADVANCED MATHEMATICS IN THE TRACTS OF MATHEMATICAL PHILOSOPHY: A BIBLIOGRAPHICAL SURVEY

In this chapter we will review the reception that advanced mathematics has enjoyed in mathematical philosophy. As we shall see, the *absences* clearly outnumber the presences, though there have been significant efforts to be open to modern and contemporary mathematics. This chapter seeks only to carve out a few bibliographic footholds in a *global* descriptive landscape. In part 3 of this essay, we will come back to several of the authors mentioned here, concentrating on far more specific and constrained *local* problematics.

The first section, 'The Place of Lautman', tries to sum up some of the main contributions made by Albert Lautman's work, as a paradigm for a philosophical perspective attentive to modern mathematics. The second section, 'Approaching Real Mathematics', surveys, in chronological order, a modest number of appearances of advanced mathematics in philosophy since Lautman, and the even slighter presence there of contemporary mathematics. The third section, 'More Philosophy, Less Mathematics', explores the spectrum of modern and contemporary mathematics that appears in certain tracts of analytic philosophy of mathematics, and that is summed up in the *Oxford Handbook of Philosophy of Mathematics and Logic*, where a deepening of philosophical aspects takes precedence over the advanced mathematics themselves, which are neglected. This is, of course, a valid *option*, but one that must be acknowledged as such: a *choice* that results in a vast panorama being disregarded.

The heart of mathematics has its reasons of which linguistic reason knows nothing. Just like Stephen, in Joyce's *A Portrait of the Artist as a Young Man*, who must confront the 'wild heart of life' and must choose whether to avoid it or rather to immerse himself in it, the philosopher of mathematics cannot avoid having to confront the 'wild heart' of mathematics. She may elude it, if that is what she wants, but her reflections would then fail to take stock of many of mathematics' central aspects; in particular, an understanding of mathematical *creativity* that neglects advanced mathematics cannot be anything other than a limited and skeletal understanding. What would we say of a historian or philosopher of art who took it upon himself to elaborate meticulous chromatic distinctions while solely restricting himself to a set of mediocre paintings, leaving aside, for example, the complex colorations of Turner, Monet or Rothko? What would we make of a literary critic who presumed to circumscribe the 'whole' of literary invention while reducing it to the short story or novella, without, for example, taking Proust or Musil into consideration?

In the arts it would be unthinkable – almost atrocious – to neglect the great creations of the genre while aspiring to elaborate an aesthetics. In the philosophy of mathematics, however, it has apparently proved rather easy to skip over the emblematic creations of advanced mathematics. In the introduction and chapter 1, we have indicated a few of the reasons why such 'oblivion' has been so very comfortable and so rarely disquieting: the belief that contemplating the world of elementary mathematics *equals* contemplating the world of advanced mathematics; the standardization of perspectives based on philosophies of language, the assumption that perceiving modern and contemporary technical advances would not lead to major changes in the philosophy of mathematics. As we understand it, this has to do with preemptively taking up positions in such a way that access to the 'real' world of mathematics, in its continuous and contemporary development, is impeded. Given that, as we have indicated in chapter 1, advanced mathematics hinges on peculiar *specificities* that distinguish it from elementary mathematics, to limit mathematical philosophy to *elementary* mathematics – however philosophically *sophisticated* our reasons for doing so might be – presupposes a dubious reductionism. We will take up, in the first two sections of this chapter, the works of a few philosophers of mathematics who have indeed attempted to access the 'wild heart' of the discipline.

2.1 THE PLACE OF LAUTMAN

Albert Lautman perhaps came closer than any other twentieth-century philosopher to understanding the creative world of modern mathematics. The 'Essay on Notions of Structure and Existence in Mathematics'[22] is the principal thesis for a doctorate in letters (philosophy), defended by Lautman at the Sorbonne in 1937. The work, dedicated to the memory of his friend and mentor Herbrand, constitutes a genuine revolution, as much in the ways of doing philosophy of mathematics, as in the depth – the undercurrent – of the ideas set down and the horizons anticipated. Lautman grafts together structural and dynamical conceptions of mathematics, interlacing the 'life' of modern mathematics with a sweeping spectrum of dialectical actions: the local and the global (chapter 1); the intrinsic and the induced (chapter 2); the becoming and the finished – closely tied to the ascent and the descent of the understanding (chapter 3); essence and existence (chapter 4); mixtures (chapter 5); the singular and the regular (chapter 6). Lautman divides his thesis into two large parts ('Schemas of Structure' and 'Schemas of Genesis') so as to emphasize one of his fundamental

---

22 A. Lautman, *Essai sur les notions de structure et d'existence en mathématiques. I. Les schémas de structure. II. Les schémas de genèse* (Paris: Hermann, 1938, 2 vols.). Republished in A. Lautman, *Essai sur l'unité des mathématiques et divers écrits* (Paris: 10/18, 1977), 21–154. Recently republished in A. Lautman, *Les mathématiques les idées et le réel physique* (Paris: Vrin,2006). [Tr. S. Duffy as *Mathematics, Ideas and the Physical Real* (London: Continuum, 2011), 87–193.]

assertions regarding the mathematics of his epoch – that modern mathematics has a structural character (a pre-figuration of the Bourbaki group – Lautman was close friends with Chevalley and Ehresmann) and consequently mathematical creativity (the genesis of objects and concepts) is interlaced with the structural decomposition of many mathematical domains.

For the first time in the history of modern mathematical philosophy, a philosopher had conducted a *sustained, profound and sweeping survey of the groundbreaking mathematics of his time*; confronting its technical aspects without ambiguity or circumlocution, and 'dividing' it into basic concepts that he painstakingly explains to the reader, Lautman presents a strikingly rich landscape of the great inventive currents of modern mathematics.[23] Thus breaking with the usual *forms* of philosophical exposition – which used to (and, unfortunately, still do) keep the philosopher at a distance from *real* mathematics – Lautman opens an extraordinary breach in an attempt to seize upon the problematics of mathematical creativity.

---

23 What follows is a brief summary of the mathematical themes reviewed by Lautman in his principal thesis. Chapter 1: complex variable, partial differential equations, differential geometry, topology, closed groups, functional approximations. Chapter 2: differential geometry, Riemannian geometry, algebraic topology. Chapter 3: Galois theory, fields of classes, algebraic topology, Riemann surfaces. Chapter 4: mathematical logic, first-order arithmetic, Herbrand fields, algebraic functions, fields of classes, representations of groups. Chapter 5: Herbrand fields, Hilbert spaces, normal families of analytic functions. Chapter 6: operators in Hilbert spaces, differential equations, modular functions. Entering into the landscape that Lautman draws, the reader is then really able to *feel* the multiple modes and creative movements of modern mathematics, never present in the elementary examples usually adduced in the philosophy of mathematics.

Breaking, too, with the common canons that are *fundamentally* accepted in the works of his contemporaries (with their epistemological or linguistic emphases), Lautman seeks to repair the complex dialectico-hermeneutic tissue that forms the backdrop of those works (with as many ties to Plato as to Heidegger), far removed from the 'naïve' Platonism that he repeatedly criticizes.

Attending to the most innovative mathematics of his time, and open to a reinterpretation of the Ideas in their original Platonic sense (as 'schemas of structure' that organize the actual), Lautman exhibits the main lines of support of mathematics' modern architectonic in his 'Essay on the Notions of Structure and of Existence in Mathematics'. Dialectical oppositions, with their partial saturations, and mixtures constructed to saturate structures, are linked to one another and to the underlying living processes of mathematical technique. The unitary interlacing of mathematical methods, through the ever-permeable membranes of the discipline's various branches, is a dynamical braiding in perpetual development. Mathematics, far from being merely one and eternal, is an indissoluble ligature of contraries: it is one-multiple and stable-evolutive. The richness of mathematics is largely due to that elastic duplicity that permits, both technically and theoretically, its *natural* transit between the ideal and the real.

In his complementary thesis for the doctorate of letters, 'Essay on the Unity of the Mathematical Sciences in their Current Development',[24] Lautman explores the profound unity of modern mathematics with more brilliant case studies. He believes he has detected that unity in the burgeoning infiltration of algebra's structural and finitary methods into all the domains of mathematics: dimensional decompositions in the resolution of integral equations (chapter 1), non-euclidean metrics and discontinuous groups in the theory of analytic functions (chapter 2), methods from noncommutative algebra in differential equations (chapter 3), modular groups in the theory of automorphic functions (chapter 4). Lautman thereby stresses the tense union of the continuous-discontinuous dialectic within modern mathematics, a union that takes on aspects of 'imitation' or 'expression' as it alternates between finite and infinite structures: imitation when, in order to resolve a problem, one seeks to trace a simple property of finite structures in the infinite; expression when the emergence of a new infinite construction includes a representation of the finite domains that prompted its emergence. The 'analogies of structure of reciprocal adaptations' between the continuous and the discontinuous are, for Lautman, among the basic engines of mathematical creativity, a fact that seems to have been

---

24 Lautman, *Essai sur l'unité des mathématiques et divers écrits*, 155–202. *Les mathématiques les idées et le réel physique* 81–124; translation, 45–83.

borne out by the entire second half of the twentieth century, both in the unitary entanglement of the methods of algebraic geometry that brought about the demonstration of Fermat's Theorem (Wiles), and in the geometrical and topological adaptations that appear to be on the verge of proving the Poincaré conjecture (Perelman).

If, in his complementary thesis, Lautman's emphasis primarily tends in the discrete $\longrightarrow$ continuous direction (the direction in which the tools of modern algebra help to generate the concepts and constructions of analysis), a study of the other direction, continuous $\longrightarrow$ discrete, can also be found in his reflections on analytical number theory in 'New Investigations into the Dialectical Structure of Mathematics'.[25] In this brief pamphlet – the last work Lautman published in his lifetime – he remarks on how 'reflections on Plato and Heidegger' are conjugated with 'observations on the law of quadratic reciprocity and the distribution of prime numbers', in an effort to sustain, once again, one of his fundamental theses: to show 'that that rapprochement between metaphysics and mathematics is not contingent but necessary'. In the Heideggerian transit between pre-ontological understanding and ontic existence, Lautman finds, in channels internal to philosophy, an important echo of his own reflections

---

25   A. Lautman, *Nouvelles recherches sur la structure dialectique des mathématiques* (Paris: Hermann, 1939). Republished in Lautman, *Essai sur l'unité...*, 203–229. *Les mathématiques...*, 125–221; translation, 197–219.

regarding the transit of the structural and the existential within modern mathematics.

In the second subsection of the first part of 'New Investigations', 'The Genesis of Mathematics out of the Dialectic', Lautman explicitly defines some of the fundamental terms that, in his thesis, are audible only in whispers: the pairs of dialectical *notions* (whole/part, extrinsic/intrinsic, system/model, etc.), and the associated dialectical *ideas*, which should be understood as partial resolutions of oppositions between 'notions'. For example, the understanding of the continuum as a saturation of the discrete (the Cantorian completion of the real line) is a Lautmanian 'idea' that partially responds to the pair of 'notions', continuous/discrete, but it is clear that there can likewise be many other alternative 'ideas' for delimiting the notions at stake (like Brouwer's primordial continuum, from which the discrete is detached, reversing Cantor's process). In virtue of Lautman's synthetic perception, mathematics exhibits all of its liveliness, and the nonreductive richness of its technical, conceptual and philosophical movements becomes obvious. And so the *harmonious* concord of the Plural and the One, perhaps the greatest of mathematics' 'miracles' shines forth.

The war years slowed the tempo of Lautman's writing – he was deeply involved in military activities and in the Resistance – but he nevertheless found time to try to lift his work above the horror that surrounded him. In 1939, during a memorable session of the French Philosophy Society, Lautman defended, alongside Jean Cavaillès, the theses the two friends had recently submitted. A transcript of Lautman's intervention is preserved in 'Mathematical Thought',[26] where he insists on the structural character of modern mathematics, and points to how contrary 'notions' (local/global, form/matter, container/contained, etc.) dwell within groups, number fields, Riemann surfaces and many other constructions, and how 'the contraries are not opposed to one another, but, rather, are capable of composing with one another so as to constitute those mixtures we call *Mathematica*'. At the end of his intervention, in an homage to Plato and a return to the *Timaeus*, Lautman proposes an ambitious reconstruction of the 'theory of Ideas' for mathematical philosophy, in three great stages: a description of the inexhaustible richness of effective mathematics; a hierarchization of mathematical geneses; a structural explanation of mathematics' applicability to the sensible universe.

Lautman's last two works take up, in part, this final task and thereby serve to bring his philosophical labor to a coherent closure. Chapters from a monograph on the philosophy of physics that Lautman was unable to complete, 'Symmetry and Dissymmetry in Mathematics and in Physics' and 'The Problem of Time', aggressively tackle the

---

26 A. Lautman, 'La pensée mathématique', *Bulletin de la Société Française de Philosophie* XL, 1946: 3–17.

problem of interlacing the ideal with the real, by way of the complex conceptual constructions of quantum mechanics, statistical mechanics and general relativity.[27] Lautman establishes some remarkable correlations 'between dissymmetrical symmetry in the sensible universe and antisymmetric duality in the world of mathematics', discovers the potential of the recently developed lattice theory (Birkhoff, Von Neumann, Glivenko) and meticulously shows that any conception of time must simultaneously account for both the global form of the entire universe and its local evolutions. Seeking an explanation of time's sensible duality (as both an oriented dimension and a factor of evolution), Lautman uncovers the duality's 'ideal' roots in a profound and original structural investigation of time's twofold behavior in differential equations. Raising mathematics and physics to the point where they are seen as higher-order 'notions' linked to symmetry (predominantly mathematical) and dissymmetry (predominantly physical), Lautman thus successfully completes the first circumnavigation of his theory of 'ideas'.

Among Lautman's many contributions, both his studies of the *mixtures* of modern mathematics and his explication of the *ideas/notions* that facilitate the incarnation of obstructions and resolutions in mathematical creativity are

---

27 A. Lautman, *Symétrie et dissymétrie en mathématiques et en physique* (Paris: Hermann, 1946); Republished in Lautman, *Essai sur l'unité...*, 231–80. *Les mathématiques...* 265–99; translation 229–62.

of particular relevance. Mathematical mixtures are legion; the ones most studied by Lautman include algebraic topology, differential geometry, algebraic geometry, and analytical number theory. These interlacings of a noun (the main subdiscipline) with an adjective (the 'infiltrating' subdiscipline) are just faint echoes of the real procedural osmoses of modern mathematics; the rigid delimitations of the past fade away, to emerge again as pliable folds within a new classification that resembles not so much a tree as a vast liquid surface over which information flows between mobile nuclei of knowledge. Albert Lautman is the *only* philosopher of modern mathematics who has adequately emphasized and studied the advent of mathematical mixtures *in actuality*, on the one hand, and the 'ideas' and 'notions' that allow the transit of those mixtures to be understood *in potentiality*, on the other. Given the inescapable importance of mixed constructions in contemporary mathematics, it is of no surprise to say that Lautman's philosophy of mathematics would be of great worth to our time, if only it were better known. To break with an imagined 'tautological' mathematics, the dubious 'pure' invention of analytic philosophy, and to open onto a *contaminated*, and much more real, mathematics, is the order of the day.

Lautman exalts the richness won by the introduction of 'transcendent' analytical methods into number theory, and explains *why* mathematical creativity has a natural tendency to require such mixtures and mediations:

> the demonstration of certain results pertaining to the integers depends upon the properties of certain analytic functions, *because* the structure of the analytical means employed turns out to be already in accordance with the structure of the sought after arithmetical results.[28]

In fact, the great interest of the mixtures lies in their capacity to partially reflect properties from one extreme to another, and to serve as relays in the transmission of information.[29] Whether they occur in a given structure (Hilbert space), in a collection of structures (Herbrand ascending domains) or in a family of functions (Montel normal families), mixtures, on the one hand, imitate the

---

28 Lautman, *Les mathématiques...* 245–6; translation, 208 (translation modified).

29 Francastel's *relé* (from the French *relais*, meaning 'relay') affords – for the work of art – another mixture or junction of great value; one where the perceived, the real and the imaginary are conjugated. 'The plastic sign, by being a place where elements proceeding from these three categories encounter and interfere with one another, is neither expressive (imaginary and individual) nor representative (real and imaginary), but also figurative (unity of the laws of the brain's optical activity and those of the techniques of elaboration of the sign as such).' (P. Francastel, *La realidad figurative* [1965] [Barcelona: Paidós, 1988], 115). If we contrast a definition of the work of art as a 'form that signifies itself' (Focillon) with a definition of the work of mathematics as a 'structure that forms itself' (our extrapolation, motivated by Lautman), we can have some intuition, once again, of the deep common ground underlying aesthetics and mathematics. For a remarkable recuperation of a history of art that takes stock of the complex and the differential, but that recomposes it all in a stratified and hierarchical dialogue, attentive to the universal and to 'truth' (an eminently Lautmanian task), see J. Thuiller, *Théorie générale de l'histoire de l'art* (Paris: Odile Jacob, 2003). For Focillon's definition, and an extensive subsequent discussion, see p. 65.

structure of underlying domains and, on the other, serve as partial building blocks for the structuration of higher domains. Without this kind of *sought-after* contamination, or premeditated alloying, contemporary mathematics would be unthinkable. A remarkable result like the proof of Fermat's Theorem (1994) was only possible as the final exertion in a complex back-and-forth in which an entire class of mathematical mixtures had intervened – a problem concerning elliptical curves and modular forms resolved by means of exhaustive interlacings between algebraic geometry and complex analysis, involving the zeta functions and their Galois representations.

'Mixtures' show up in Lautman's earliest surviving manuscript – his posthumously published work on mathematical logic.[30] At twenty-six, Lautman describes the construction of Herbrand 'domains' and shows that 'the Hilbertians succeeded in interposing an intermediary schematic of individuals and fields considered not so much for themselves as for the infinite consequences that finitary calculations performed on them would permit'.[31] Comparing that 'intermediary schematic' with Russell's hierarchy of types and orders, Lautman points out that 'in both cases, we are faced with a structure whose elements are neither entirely arbitrary nor really constructed,

---

30 A. Lautman, 'Considérations sur la logique mathématique', in Lautman, *Essai sur l'unité...*, 305–15. Republished in Lautman, *Les mathématiques...* , 39–46; translation 1–8.

31 Ibid., 315; 46; translation, 7–8 (translation modified).

but composed as a *mixed form* whose fertility is due to its double nature'.[32] This clear intuition of the mixed forms of logic, in the thirties, when logic, by contrast, tended to see itself as a 'pure form', is evidence of the young philosopher's independence and acumen. Indeed, it now seems obvious that those mixed forms of logic were the underlying cause of the late twentieth-century blossoming of mathematical logic, happily infiltrated by algebraic, topological and geometrical methods. In that sense, Lautman never presupposes an a priori logic, prior to mathematics, but considers it as a constituent part of doing mathematics, with a prescient sense of today's pluralistic conception of logic, whereby a logical system *accommodates* a collection of mathematical structures rather than installing itself beneath it.

In a text on the 'method of division' in modern axiomatics (his first published article), Lautman already moves to link the mention of mixtures with the great philosophical tradition:

It is not Aristotelian logic, the logic of genera and species, that intervenes here [i.e., in mathematical creation], but the Platonic method of division, as it is taught in the *Sophist* and the *Philebus*, for which the unity of Being is a unity of composition and a point

---

32 Ibid, 315; 46; translation, 8 (translation modified).

of departure for the search for the principles that are united in the ideas.[33]

Lautman accentuates the dynamical interest of a mixture, 'which tends to liberate the simple notions in which that mixture participates', and thereby situates mathematical creativity in a dialectic of *liberation* and *composition*. In terms foreign to Lautman, but which situate his position on better-known terrain, on the one hand, the making of mathematics divides a concept's content into definitions (syntax) and derivations (grammar), and liberates its simple components; on the other hand, with models (semantics) and transfers (pragmatics), it constructs intermediary entities that resume the existence of those simple threads, recomposing them into new concepts. When the mixture succeeds in simultaneously combining a great simplicity and a strong power of reflection in its components – as with Riemann surfaces or Hilbert spaces, the subject of admiring and exemplary studies by Lautman – mathematical creation reaches perhaps its greatest heights.

In Lautman's theses, the philosopher's entire reflective movement is propelled by a pendular contrasting of complementary concepts (local/global, whole/part,

---

33 A. Lautman, 'L'axiomatique et la méthode de division', in *Recherches philosophiques* VI, 1936–37: 191–203; republished in Lautman, *Essai sur l'unité...*, 291–304. *Les mathématiques...*, 69–80; translated in Lautman, *Mathematics, Ideas...*, 31–42.

extrinsic/intrinsic, continuous/discrete, etc.), but it is in his 'New Investigations into the Dialectical Structure of Mathematics' that Lautman introduces the terms that govern those dialectical interlacings. Lautman defines a *notion* as one of the poles of a conceptual tension and an *idea* as a partial resolution of that polarity. The concepts of finitude, infinity, localization, globalization, calculation, modeling, continuity and discontinuity (Lautman's examples) are Lautmanian 'notions'. A few Lautmanian 'ideas' (our examples) would be the proposals according to which the infinite is grasped as nonfinite (the cardinal skeleton), the global as gluing of the local (compactness), the model-theoretic as realization of the calculative (set-theoretical semantics), or the continuous as completion of the discrete (the Cantorian line).

The interest of 'notions' and 'ideas' is threefold: they allow us to *filter out* (liberate) unnecessary ornaments and decant the grounds of certain mathematical frameworks; they allow us to *unify* various apparently disparate constructions from the perspective of a 'higher' problematic level; and they allow us to *open* the mathematical spectrum to various options. Whether the mathematical landscape is being filtered or unified (duality theorems in algebraic topology and lattice theory) or being opened towards a fuller scope of possibilities ('non-standard ideas', which resolve *in another way* oppositions between fundamental 'notions': the infinite as the immeasurable in Robinson, the discrete as demarcation of a primordial continuum in Brouwer, the calculative as system of coordinates for the model-theoretic in Lindström), Lautmanian 'notions' and 'ideas' let us roam *transversally* over the universe of mathematics and explicate not only the *breadth* of that universe, but its harmonious concord between the One and the Many.

For Lautman, 'notions' and 'ideas' are situated on a 'higher' level, where the intellect is capable of imagining the *possibility of a problematic*, which, nevertheless, acquires its sense only through its immediate incarnation in real mathematics. Lautman is conscious of how an a priori thereby seems to be introduced into the philosophy of mathematics, but he explains this as a mere 'urgency of problems, prior to the discovery of solutions'. In fact, a problem's 'priority' itself should be only considered as such from a purely conceptual point of view, since, as Lautman himself points out, the elements of a solution are often found to have already been given *in practice*, only *later* to incite the posing of a problem that incorporates those data (none of which prevents, in a conceptual reordering, the problem from ultimately preceding the solution). In parallel with his strategy of apprehending the global structure of a theory rather than predefining its logical status, Lautman consistently situates mathematical logic as an activity within mathematics that should not arbitrarily precede it and that should be situated *on the same level* as any other mathematical theory, thus anticipating the current conception of logic, which has been accepted in the wake of model theory. Following Lautman, 'logic requires a mathematics in order to exist', and it is in the weaving together of *blended* logical schemata and their effective realizations that the force of doing mathematics lies.

It is in the tension between a 'universal' (or 'generic') problematic and its 'concrete' (or 'effective') partial resolutions, according to Lautman, that the better part of the structural and unitary weaving of mathematics may take root. As we will see in chapter 7, this is precisely the paradigm proposed by the mathematical theory of categories.[34] When Lautman looks at Poincaré and Alexander's duality

---

34 Lautman never lived to know category theory, the rise of which began at the very moment of his death (S. Eilenberg, S.Mac Lane, 'Natural isomorphisms in group theory', *Proc. Nat. Acad. Sci.* 28, 1942: 537–43; S. Eilenberg, S. Mac Lane, 'General theory of natural equivalences', *Trans. Amer. Math. Soc.* 58, 1945: 231–94). It is difficult to know to what extent conversations with his friend Ehresmann – who introduced the general theory of fiber spaces in the forties and promoted category theory in France from the end of the fifties on – could have influenced, in its *implicit depths*, a conception of mathematics so clearly recognizable (in retrospect) as categorical as Lautman's is. Nevertheless, in the session of the French Philosophy Society in which Cavaillès and Lautman defended their work, and in which Ehresmann participated, the latter already pointed out precisely how a number of Lautman's philosophical conceptions should be technically filtered and converted into equipment *internal to mathematics itself*: 'If I have understood correctly, in the domain of a supramathematical dialectic, it would not be possible to specify and investigate the nature of those relations between general ideas. The philosopher could only make the urgency of the problem evident. It seems to me that if we preoccupy ourselves with speaking about those general ideas, then we are already, in a vague way, conceiving of the existence of certain relations between those general ideas. From that moment, we can't then just stop in the middle of the road; we must pose the problem, the genuinely mathematical problem, that consists in explicitly formulating those general relations between the ideas in question. I believe that a satisfactory solution can be given to that problem, regarding the relations between the whole and its parts, the global and the local, the intrinsic and the extrinsic, etc. [...] I believe that the general problems that Lautman poses can be stated in mathematical terms, and I would add that we can't avoid stating them in mathematical terms.' The entire rise of category theory effectively bears out Ehresmann's position.

theorems, and describes how 'the structural investigation of a space that receives a complex is reduced to
the structural investigation of that complex',[35] when he
analyses the ascent toward a universal covering surface
and contemplates the hierarchy of isomorphisms 'between
the fundamental groups of different covering surfaces of
a given surface F and the subgroups of the fundamental
group F', when he mentions an inversion between Gödel's
completeness theorem and Herbrand's theorem, which he
later extends to an alternation between form and matter
by way of certain mediating structures, or when – bolder
still – he asks if 'it is possible to describe, at the heart
of mathematics, a structure that would be something
like a initial sketch of the temporal form of sensible
phenomena',[36] Lautman is in each case anticipating certain
techniques in categorical thought. These include functors
in algebraic topology, representable functors in varieties, adjunctions in logic, and free allegories. In fact, by
'admitting the legitimacy of a theory of abstract structures,
independent of the objects linked to one another by those
structures', Lautman intuits a mathematics of structural
relations *beyond a mathematics of objects* – which is to say,
he prefigures the path of category theory.

---

35 Lautman, *Les mathématiques...*, 201; translation, 162 (translation modified).
36 Ibid., 173; translation 132 (translation modified).

The Lautmanian language of 'notions', 'ideas' and dialectical hierarchies finds, in category theory, a definite technical basis. 'Notions' can be specified by means of universal categorical constructions (diagrams, limits, free objects), 'ideas' by means of elevations of classes of free objects and functorial adjunctions, dialectical hierarchies by means of scales of levels of natural transformations. In this way, for example, Yoneda's Lemma technically explicates the *inevitable* presence of the ideal in any thorough consideration of mathematical reality (one of Lautman's basic contentions), showing that *every* small category can be immersed in a category of functors, where, in addition to the representable functors that form a 'copy' of the small category, there also inevitably appear additional ideal functors ('presheaves'), that *complete* the universe. What is at issue here is the ubiquitous appearance of the 'ideal' whenever the capture of the 'real' is at stake, a permanent and pervasive osmosis in every form of mathematical creativity.

The majority of the schemas of structure and genesis that Lautman studies in his principal thesis can be categorically specified and, most importantly, extended. For example, the 'duality of local and global investigations' is grafted onto a complex set of instruments of functorial localizations and global reintegrations (Freyd-style representation theorems), the 'duality of extrinsic and intrinsic points of view' feeds into the power of a topos's internal logic (Lawvere-style geometrical logic), and the 'interest of the logical scheme of Galois theory' is extended into a general theory of residuality (categorical Galois connections in the style of Janelidze). So, when we see how Lautman observes that 'certain affinities of logical structure allow us to approach different mathematical theories in terms of the fact that each offers a different sketch of a solution for the same dialectical problem', that 'we can speak of the participation of distinct mathematical theories in a common Dialectic that dominates them', or that the 'indetermination of the Dialectic [...] simultaneously secures its exteriority', it seems natural to situate his ideas in a categorical context, whether in the weaving between abstract categories ('common Dialectic') and concrete categories ('distinct mathematical theories'), or in terms of free objects ('indetermination of the Dialectic') whose extensive external applicability throughout the entire spectrum of mathematics is precisely a consequence of their schematic character.

The mutual enrichment of effective Mathematics and the Dialectic (Lautman's capitalizations) is reflected in the natural ascent and descent between Lautmanian *notions and ideas*, on the one hand, and *mixtures*, on the other. In fact, in ascending from the mixtures, the 'notions' and 'ideas' that allow us to situate the place of those mixtures within an amplified dialectic are liberated; while descending from the 'notions', new mixtures are elaborated in order to specify and *incarnate* the content of the 'ideas' at stake. One of the great merits of Lautman's work consists in its having shown how those processes of ascent and descent must be *indissolubly* connected in the philosophy of mathematics in extenso, just as they are in a Galois correspondence in nuce.

2.2 APPROACHING 'REAL MATHEMATICS'

In the following pages we will perform a brief survey of the works of other authors who have tried to approach the 'heart' of 'real mathematics'. The survey will be chronological, and may be considered adequately representative, but it is certainly not exhaustive. For each work, we will indicate, firstly, what spectrum of mathematics it examines, and secondly, what global accounts obtain in light of such an examination. As we shall see, these approaches for the most part appear to concern classical mathematics (Pólya, Lakatos, Kline, Wilder, Kitcher), though other endeavors seek to examine modern (de Lorenzo, Mac Lane, Tymoczko, Châtelet, Rota), and even contemporary (Badiou, Maddy, Patras, Corfield), mathematics. We do not find a comprehension of modern mathematics as precise and broad as that achieved by Lautman in any of the cases of which we are aware.

George Pólya

Pólya's works constitute a mine of examples for bringing the reader closer to the processes of discovery *and* invention (*both* processes being indispensable) at work within classical and elementary mathematics. *Mathematics and Plausible Reasoning* presents an important collection of case studies that concern two major themes: analogical and inductive constructions in mathematics, and modes of probable inference.[37] Volume 1 (*Induction and Analogy in Mathematics*) explores classical analysis (primarily concerning the figure of Euler, gloriously resurrected), the geometry of solids, elementary number theory, the study of maxima and minima, and certain elementary problems from physics. Pólya carefully examines the weavings between generalization and specialization, certain classes of analogical hierarchies, the construction of the multiple steps of a demonstration and conjecture confirmation. He includes numerous, thoughtful exercises (with solutions), which should be seen as a means for opening the (philosophical or mathematical) reader's mind to a nondogmatic understanding of mathematical practice. Volume 2 (*Patterns of Plausible Inference*) confronts the problem of the plausibility of certain hypotheses from which one deduces further statements (a sort of inverted

---

37 G. Pólya, *Mathematics and Plausible Reasoning*, 2 vols. (Princeton: Princeton University Press, 1954).

modus ponens, corresponding to Peirce's 'retroduction', which Pólya, however, does not mention). The problem of plausible inference $[A \longrightarrow B, B \blacktriangleright\blacktriangleright A]$ consists in specifying the conditions on the deduction $A \longrightarrow B$ and on the possible truth of B that must obtain in order for the retroduction $\blacktriangleright\blacktriangleright$ to A to be as plausible as possible. Pólya goes on to tackle the progressive gradations of proof, the small internal variations that allow us to overcome the obstructions encountered in the solution process, inventive chance, and the back-and-forth of hypotheses and intermediate lemmata by which a demonstration continuously takes shape. Through a close inspection of classical mathematics, he detects a complex hierarchization, which was later exploited in modern mathematics. The other modern characteristics (semantic richness, theorematic mixtures, structural unity) nevertheless fail to appear within the classical horizon.

In *Mathematical Discovery*, Pólya restricts himself to examples from elementary mathematics (basic geometrical forms, numerical sums, Pascal's triangle) – though he also includes a few classical references related to limits and power series – to illustrate the gradual emergence and concretization of mathematical ideas, from the vague and apparently contradictory to the measured control of a proof.[38] By means of processes of figuration, superposition

---

38 G. Pólya, *Mathematical Discovery* (New York: Wiley, 1962).

and amplification, Pólya shows how webs of auxiliary notions and problems gradually take shape, converging toward the solution of an initial problem, and how a surprising interlacing of chance and discipline is often found lurking behind various demonstrations. Numerous examples and exercises with solutions once again bring the reader closer to mathematical practice. This practical approach offers us a sense, however faint, of modern mathematics' dynamical richness.

Imre Lakatos

Lakatos systematically introduces, into the philosophy of mathematics, the method of conjectures and refutations that Popper had applied to the philosophy of science as a whole. In *Proofs and Refutations: The Logic of Mathematical Discovery*, Lakatos explores the fluctuating mechanisms of mathematical discovery, the changing norms of proofs, the interlacing of counterexamples and lemmata in the construction of a demonstration and the back-and-forth of a mathematics understood as an experimental science.[39] The examples adduced are eminently classical and are treated with patience and care: Euler's Polyhedron Theorem, Cauchy and the problems of uniform convergence, bounded variation in the Riemann integral. Many dialectical

---

39 I. Lakatos, *Proofs and Refutations: The Logic of Mathematical Discovery* (Cambridge: Cambridge University Press, 1976). The book extends earlier articles from 1963-64.

forms – explicit uses of the thesis-antithesis-synthesis triad, expository games in Platonic dialogue, an incessant back-and-forth between obstructions and resolution – run through the work, which thereby discerns the emergence of the dialectical dynamic that will govern the development of modern mathematics.

Mathematics, Science and Epistemology posthumously assembles Lakatos's articles on the philosophy of mathematics.[40] The spectrum observed is once again the environment of classical mathematics (from the Greeks up to Abel and Cauchy, on whom Lakatos's considerations are centered), but the book also contains various commentaries on the modern foundations (Russell, Tarski, Gödel), following the line preponderantly adopted by twentieth-century philosophy of mathematics. Contemporary mathematics makes an appearance only in the form of Robinson's nonstandard analysis, as a result of Lakatos's interest in connecting it with a recuperation of the infinitesimals utilized by Cauchy. Various hierarchies are proposed, dealing with the steps of a proof (preformal, formal, postformal), and examples dealing with the method of conjectures, proofs and refutations are refined. The profound instruments of algebraic geometry which, by the sixties, had already been constructed (by Grothendieck), and which later led to proofs of the great theorems

---

40 I. Lakatos, *Mathematics, Science and Epistemology*; vol 2 of *Philosophical Papers* (Cambridge: Cambridge University Press, 1978).

of arithmetic, such as the Weil conjectures (Deligne, 1973), go unmentioned. Instead, somewhat dubious speculations are entertained regarding the undecidability of Fermat's Theorem – yet another example of the *distance* between the philosopher of mathematics and the mathematics of *his* epoch, preventing him from taking anything like *snapshots* of the mathematical thinking being forged around him.

Javier de Lorenzo

In his first monograph, *Introducción al estilo matemático* [*Introduction to Mathematical Style*], De Lorenzo immediately shows himself to be awake to the modes of 'doing' advanced mathematics.[41] With creative verve, the author confronts the great figures of modern mathematics (Cauchy, Abel, Galois, Jacobi, Poincaré, Hilbert, the Bourbaki group, etc.) and argues that certain *fragments* of advanced mathematics – group theory, real analysis, and abstract geometries are his preferred examples – bring with them *distinct* ways of seeing, of intuition, of handling operations and even distinct methods of deduction, in each of their conceptual, practical and formal contexts. De Lorenzo points out how mathematics 'grows through contradistinction, dialectically and not organically', and thereby breaks with a traditional vision of mathematics, according

---

41 J. de Lorenzo, *Introducción al estilo matemático* (Madrid: Tecnos, 1971).

to which it grows by accumulation and progress in a vertical ascent. He proposes instead a conceptual amplification of the discipline, in which new realms interlace with one another *horizontally*, without having to be situated one on top of the other.

In *La matemática y el problema de su historia*, De Lorenzo postulates a radical historicity of doing mathematics.[42] The references to advanced mathematics are classified in terms of three primary environments, within which, according to De Lorenzo, the major ruptures and inversions that gave rise to modern mathematics were forged: *the environment of 1827*, in which the program for the resolution of mathematical problems is inverted, setting out 'from what seems most elusive in order to account for why [problems] can or cannot be resolved', and in which mathematics begins to feed on itself and its own limitations; *the environment of 1875*, in which the mathematical tasks of the previous half century are unified (groups, sets) or transfused from one register to another (geometrical methods converted into algebraic or axiomatic methods), generating the important constructions (Lie groups, point-set topology, algebraic geometry, etc.) that drove mathematics' development at the outset of the twentieth century; *the environment of 1939*, in which the Bourbaki group fixed the orientation of contemporary mathematics

42 J. de Lorenzo, *La matemática y el problema de su historia* (Madrid: Tecnos, 1977).

around the notions of structure and morphism, inverted the focus of mathematical research, and moved toward a primordial search for relations between abstract structures (algebras, topologies, orders, etc.). In this and other works (see note 13), De Lorenzo also exhibits a subtle attention to contemporary mathematics (with detailed citations of Weil, Schwartz and Lawvere, for example), though modern mathematics remains his primary focus.[43] To sum up, De Lorenzo argues that mathematical knowledge is produced through very different *contexts* and branches, following many tempos and rhythms. Incessant incorporations, *transfers*, osmoses, translations and representations are afterward produced between the various environments of mathematical knowledge; already-constructed notions then give rise to new constructions by means of diverse deformations and *transfigurations*.

---

43 Regarding Lawvere, for example, De Lorenzo points out – only *seven* years (!) after Lawvere introduced elementary topoi (1970) – that 'the interlacing of the theory of categories with that of topoi, presheaves and algebraic geometry is showing itself to be essential for the intentions of Lawvere and those working in the same direction, to achieve a foundation, which he qualifies as "dialectical", for mathematical work, while recognizing that such a foundation can only be of a descriptive character, achieving in this way a revision of Heyting's intuitionistic logic as the one best adapted to topos theory.' The investigation of the mathematical *in progress* ('*is showing itself to be...*', '*those working...*') not only surfaces in these unusual meditations of a historian and philosopher, but is made in the *most fitting* possible way, successfully detecting the *conceptual kernel* of the situation: the interlacing of topoi with algebraic geometry and with the underlying intuitionistic logic.

Raymond L. Wilder

Mathematics as a Cultural System puts forward a valuable and original conception of mathematics as a 'vectorial system' in which various tendencies of mathematics counterpose, superpose, interlace and consolidate with one another, as if they were situated in a web of vectorial operations.[44] Rather than understanding the realm of mathematics according to the dispersive model of a 'tree', the vectorial system permits the introduction, with greater finesse, of the fundamental ideas of directionality, potentiality, normalization and singularity associated with vector fields. Wilder explores many examples in classical mathematics (Leibniz, Fermat, Gauss and, in particular, Desargues), and in modern mathematics (Bolzano, Lobachevski, Riemann, Hilbert), where a living dialectic is established between potential fields (e.g., the resolution of algebraic equations), normal vectors (e.g., ad hoc manipulations by radicals) and singularities (e.g., the 'ingenious' emergence of Galois). Wilder's great knowledge of modern topology and algebra – he is one of the few *active* mathematicians (together with Pólya and Mac Lane) to appear in this chapter's bibliographical survey – allows for a detailed demonstration of the fact that mathematical 'reality' is a sort of changing

---

44   R. Wilder, *Mathematics as a Cultural System* (Oxford: Pergamon Press, 1981).

flux in the conceptual field of associated vectors, and that various tendencies undergo constant modification in accordance with their historical position in the web. An evolution of *collective* mathematical intuition and a search for *invariants* in that evolution allow us to see how mathematical knowledge naturally modifies and stabilizes itself, perduring despite its own plasticity.

Morris Kline

In *Mathematics: The Loss of Certainty*, the vision of a great connoisseur of the history of mathematics takes flight, accompanied by philosophical speculations that, however, are quite a bit weaker.[45] Kline proves to be particularly attentive to four principal registers: Greek mathematics; classical analysis (rise, development, disorder, foundations, crisis, limitations); modern mathematics (reviewing various works of Poincaré, Weyl, Borel, Hilbert, von Neumann, Stone, Dieudonné, etc.); the foundations of mathematics (Cantor, Brouwer, Gödel, etc.). Curiously, for all his profound and extensive historical knowledge, the reflections that such knowledge gives rise to are debatable to say the least: an insistence on an 'illogical' development of mathematics (in which errors, conceptual shiftings and the recourse to intuition play a leading role),

---

45 M. Kline, *Mathematics: The Loss of Certainty* (New York: Oxford University Press, 1980).

the perception of an 'unsatisfactory' state of mathematics, the proclamation of an 'end to the Age of Reason', the sense of a shattered multiplicity of mathematics with no possibility for unification and the indication of a growing isolation bringing about 'disasters' in the discipline. It is surprising to find such a negative vision of mathematics take shape in the 1980s, when the discipline found itself in full bloom. Once again, the *preemptive* occupation of a philosophical position – Kline's postmodern predilection for the supposed 'loss of certainty' – clouds the vision and obscures the dynamic technical life that presents itself all around the observer. If some of the critical points are valuable (the place of error, multiplicity, relativity), to carry them to the extreme and separate them from their natural polar counterparts (proof, unity, universality) brings about an excessive oscillation of the pendulum, which impedes any detection of a far more complex relational warp.

Philip Kitcher

The Nature of Mathematical Knowledge continues the focus on episodes of classical mathematics in the tradition of Pólya and Lakatos.[46] The examples investigated include Newton, Leibniz, Bernoulli, Euler, Cauchy, and an

---

46   P. Kitcher, *The Nature of Mathematical Knowledge* (New York: Oxford University Press, 1983).

extensive case study (chapter. 10) reviews the development of analysis (1650–1870). Modern mathematics shows up much more intermittently, and the 'elementary' references to Galois (reduced to the problem of the insolubility of equations) and Riemann (with respect to the construction of his integral) are symptomatic in that sense. Focused fully on the classical spectrum, several of Kitcher's reflections prefigure with great acumen the complex web of ideal constructions and operations that will appear in modern mathematics, as well as its incessant evolution and its coupling between conceptual fragments and real data. Particularly sensitive to mathematical *change*, Kitcher succeeds in evoking the dynamism of mathematics and the discipline's unpredictable transit between the ideal and the real as well as between the possible, the actual and the necessary.

Thomas Tymoczko

Tymoczko's work as editor of *New Directions in the Philosophy of Mathematics* helps to clearly explicate the two great cascades into which the philosophy of mathematics might flow, following analytic philosophy's many decades of dominance.[47] The first part of the book ('Challenging Foundations') reminds us that philosophy has many

---

47   T. Tymoczko, ed., *New Directions in the Philosophy of Mathematics* (Boston: Birkhäuser, 1986).

other themes to study within mathematics, aside from foundations.[48] The second part ('Mathematical Practice') points out that the philosopher should also be inclined to observe mathematical practice, the evolutions of standards such as 'truth' and 'proof', the oscillation between informal and rigorous proofs, and the complexity of the mathematical architectonic. It is in the second part that the articles coming closest to modern and contemporary mathematics appear (Tymoczko on the four-color problem; Chaitin on computational complexity). But the texts for the most part continue to evoke classical examples (as Grabiner does, with respect to the development of analysis in the eighteenth and nineteenth centuries).[49] The sort of 'quasi-empiricism' that Tymoczko adopts indicates that a deeper knowledge of mathematical practice could help to resolve certain philosophical controversies regarding realism and idealism, and that therefore (and

---

48 Paul Bernays, one of the great champions of the foundations of mathematics, already pointed out *in 1940*, in a little-known review of Lautman's works, that 'it is to be said in favor of Lautman's method that it is more suited than foundational discussions to give to a philosopher an impression of the content and nature of modern mathematics. Indeed it is worthwhile to emphasize that foundational problems by no means constitute the only philosophically important aspect of mathematics' (P. Bernays, 'Reviews of Albert Lautman', *Journal of Symbolic Logic* 5, 1940]: 22). This admirable display of conscience by a genuine *architect* of the foundations of mathematics is something of which too many *philosophers* of the foundations have stood in need.

49 The inclusion of an 'interlude' with two of Pólya's texts – written thirty years earlier – is indicative of the meekness that has emerged in philosophy with respect to approaching 'mathematical practice'. Of course, as is often the case in the Anglo-Saxon academy, there is an obvious ignorance of anything not translated into English: to speak of mathematical 'practice' without mentioning Lautman or De Lorenzo is genuinely misguided, notwithstanding the ease with which this is done by anglophone philosophers.

this is one of the central foci of the present work) not only more philosophy, but *more mathematics*, could be of great assistance in resolving certain quandaries in mathematical philosophy.

Saunders Mac Lane

Mathematics: Form and Function synopsizes the perspective of an outstanding mathematician of the second half of the twentieth century.[50] The main part of the monograph – which should be seen more as a presentation, a bird's eye view, of classical and modern mathematics, than as a volume of mathematical philosophy – confronts head-on the legacy of Galois and Riemann, and provides excellent introductions to central themes in mathematics: groups, algebraic structures, complex analysis, topology. Contemporary mathematics appears with respect to category theory (Mac Lane was one of its founders) and sheaf theory (a paradigm of contemporary methods). Chapter 12, 'The Mathematical Network', explores the progressive emergence of mathematical constructions (origins, ideas, formal versions), and the incessant back-and-forth between, on the one hand, themes, specialties, and subdivisions of mathematical knowledge, and, on the other, transits, transformations and changes. For Mac Lane,

---

50 S. Mac Lane, *Mathematics: Form and Function* (New York: Springer, 1986).

mathematical constructions arise by virtue of a *network* of analogies, examples, proofs and shifts in perspective, which let us encounter and define certain *invariants* amid the change. If there is not an absolute 'truth', external to the network, there nevertheless exist multiple *gradations of relevance*, of correctness, approximation and illumination *inside* the network. It has become one of mathematics' central tasks to achieve the harmonious concord of those gradations, to overcome multiple obstructions, to construct new concepts with the residues.

Gian-Carlo Rota

Indiscrete Thoughts consists of an irreverent series of reflections, of great interest,[51] by another leading mathematician of the second half of the twentieth century.[52] It is an uneven compilation, which includes anecdotes, historical fragments, mathematical and philosophical reflections, critical notes and brilliant, incendiary ideas. Above all, and in the order of the compilation itself, Rota dedicates a great deal of space to the biographies of mathematicians (Artin, Lefschetz, Jacob Schwartz, Ulam) as creative

---

51 I am grateful here to the teachings of Alejandro Martín and Andrés Villaveces, who explained to me one memorable afternoon the importance of Rota's ideas, several of which we will return to (by different routes) in part 3 of this book. F. Palombi, *La stella e l'intero. La ricerca di Gian-Carlo Rota tra matematica e fenomenologia* (Torino: Boringhieri, 2003) presents several ideas of utmost relevance to our focus, and upon on which we will later comment.

52 G.-C. Rota, *Indiscrete Thoughts* (Basel: Birkhäuser, 1997).

individuals. For Rota, mathematics emerges in very specific vital and academic contexts (see the beautiful text on 'The Lost Café'), giving rise to a dynamic, oscillating, fluctuating discipline, with multiple concrete tensions, indissolubly bound to personalities firmly situated in place and time. The weaving between a generic mathematics and its particular incarnations, and the idea according to which 'mathematics is nothing if not a historical subject *par excellence*' (something that Jean Cavaillès had forcefully underscored half a century earlier), underlie the whole of Rota's thought and permeate some of his most original conceptions: a 'primacy of identity', which aims to define the 'essence' of an object as its very web of factual superpositions, and which would help to replace an obsolete mathematical ontology (the 'comedy of existence' of mathematical objects); a reappropriation of the Husserlian notion of *Fundierung* (*founding*) in order to rethink the mathematical transits between the factual and the functional; a phenomenology of mathematics open to forms of doing mathematics (beauty, varieties of proof, imagination) usually neglected by the traditional perspectives of mathematical philosophy.

The caustic and polemical article 'The Pernicious Influence of Mathematics upon Philosophy' reveals the excesses of a philosophy of mathematics oriented toward formal juggling acts and bastardized by various 'myths' that have little to do with mathematical practice:

the illusion of precision, axiomatic absolutism, the illusion of permanence, conceptual reducibility. Paradoxical as it may seem, Rota observes that analytic philosophy, 'perniciously influenced' by classical logic and by set theory, has *turned its back on* and has abandoned high mathematical creativity, be it geometrical, topological, differential, algebraic or combinatorial, thereby estranging itself from the real center of the discipline that helped it to emerge. The philosophy of mathematics should therefore turn back to examine, without prejudices and without taking preestablished theoretical positions, the *phenomenological spectrum of mathematical activity*. Here, Rota's reading – in three central articles on 'The Phenomenology of Mathematical Truth', 'The Phenomenology of Mathematical Beauty', 'The Phenomenology of Mathematical Proof', and, in four complementary texts, 'The Primacy of Identity', '*Fundierung* as a Logical Concept', 'Kant and Husserl', and 'The Barber of Seville or the Useless Precaution' – poses some vital problematics to which a philosophy of mathematics aiming at a 'real' (in Corfield's sense) understanding of the discipline should be open. These include the emergence of mathematical creativity, mathematics understood as the history of its problems, the varieties of proof and the evolution of concepts, the interlacings between the 'facts' of mathematics and their constant functional reinterpretations, the superpositions and nonreductive iterations of mathematical objects, and the ubiquitous transits between forms of analysis and forms of synthesis. Rota's style – brief, distilled, caustic – is not conducive to a systematic elaboration of his ideas, but we will develop a few of them in part 2.

Alain Badiou

Being and Event[53] offers a sophisticated example of how to construct *new* philosophical meditations on the basis of a patient observation of aspects of advanced mathematics.[54] Badiou carefully explores Cohen's technique of *forcing* – going beyond the mathematicians themselves in the profundity and originality of his analysis – and encounters one of the great contemporary supports by means of which the Many and the Oone may be soundly reintegrated. The investigation of the continuum hypothesis, with its contrast between indiscernibility (Easton's Theorem) and linguistic control (Gödel's constructible universe), exhibits certain oscillations of mathematical thought in fine detail. A profound *ontological subversion* is suggested – the identification of 'mathematics' (the science of pure multiplicities) and 'ontology' (the science

---

53   A. Badiou, *L'être et l'événement* (Paris: Seuil, 1988). [Tr. O. Feltham as *Being and Event* (London: Continuum, 2005).]

54   Badiou explicitly declares himself Lautman's admirer and heir. It is a unique case of recognition and shared labor, even if the mathematical spectrum covered by Lautman is much broader. Both Lautman and Badiou aim, however, to rethink and return to Plato, setting out from the exigencies of contemporary thought.

of what is, *insofar* as it is), in virtue of the sheer force of axiomatic set theory, which lets us name *all* the multiplicities of mathematics and develop a (hierarchical, complex, demonstrative) study of those multiplicities insofar as they 'are'. Badiou's text includes a great many 'chronicles of proofs' (the author's expression – i.e., proofs deconstructed from the formal language and reconstructed in a conceptual and philosophical language) which detail an unusually broad landscape of modern and contemporary set theory.

In his *Short Treatise on Transitory Ontology*,[55] Badiou continues his ontological 'subversion', so as to involve an incisive re-envisioning of category theory and the theory of elementary topoi. The construction of a *dialogue* between great figures of philosophy (Aristotle, Plato, Descartes, Spinoza, Leibniz, Kant), contemporary philosophers (Deleuze), poets (Mallarmé) and mathematicians both modern and contemporary (Cantor, Gödel, Cohen, Lawvere) is supremely original. We find suggestions of the primacy of 'real' mathematics and a consequent subordination of logic (topoi and *associated* logics, classes of structures and associated logics, the emergence of geometrical logic, an irreducible logical weaving between the global and the local), which should

---

55 A. Badiou, *Court traité d'ontologie transitoire* (Paris: Seuil, 1998). [Tr. N. Madarasz as *Briefings on Existence: A Short Treatise on Transitory Ontology* (Albany, NY: SUNY Press, 2006).]

bring about certain 'turns' in mathematical philosophy, beyond analytic philosophy and the philosophy of language. Badiou's nontrivial Platonic orientation (that is to say, one not reduced to the 'external' existence of mathematical ideas and objects), an orientation that accords with the 'condition of modern mathematics', is summed up in three points: Mathematics is a thought (entailing, against Wittgenstein's *Tractatus*, the existence of dynamic processes that cannot be reduced to language); mathematics, like all thought, knows how to explore its boundaries (undecidability, indiscernibility, genericity – entailing the irreducibility of mathematics to a set of intuitions or rules fixed in advance); mathematical questions of existence refer only to the intelligible consistency of the intelligible (entailing a marked indifference to 'ultimate' foundations, and the adoption, instead, of a criterion of 'maximal extension' for all that is 'compossible', quite similar to the richness of contemporary model theory). Mathematics – and ontology, with which it is identified – is thereby understood as a sophisticated sheaf of methods and constructions for the systematic exploration of the *transitory*.

Penelope Maddy

The contrast between the works of Badiou and those of Maddy could not be greater, even though both refer to the *same* mathematical spectrum: twentieth-century set theory. In *Realism in Mathematics*, Maddy explores descriptive set theory, the large cardinal axioms and the continuum hypothesis, and she performs a detailed survey of the contributions of the field's leading figures, from Borel and Lusin to Martin, Moschovakis and Solovay.[56] Maddy shows that the richness of the set-theoretical universe (new methods and models, new connections and perspectives, the possibility of obtaining verifiable consequences) allows us to uphold a certain 'realism' – close to some of Gödel's ideas – and to dismantle Benacerraf's dilemma, since the notions of causality associated with the dilemma lose their traction in set theory's sophisticated relative consistency proofs. Though Maddy finds a certain set-theoretical stability exactly where Badiou underscores continuous transition above all else, we should point out that both, specifically regarding the mathematics of *their* time, succeed in proposing *new* questions and resolutions for mathematical philosophy (the dissolution of Benacerraf's dilemma, the program for a transitory ontology). The labor of the philosopher attentive to the mathematics of her epoch is thus far from negligible.

In *Naturalism in Mathematics*, Maddy explores the status of additional axioms for set theory, from the double point of view of realism (the existence of objective

---

56   P. Maddy, *Realism in Mathematics* (Oxford: Oxford University Press, 1990).

universes of sets) and naturalism (the internal sufficiency of mathematics and set theory, without need of external justifications).[57] Maddy reviews various axioms of great mathematical interest (choice, constructibility, determinacy, measurability, supercompactness, etc.), and set theory's major modern architects (Cantor, Dedekind, Zermelo, Gödel) appear extensively in her monograph, as do some of its greatest contemporary practitioners (Cohen, Martin, Moschovakis, Woodin, etc.). An emphatic observation of *practice* runs through the entire text; a naturalist vision of set theory is sustained through the direct contemplation of how the set theoretical axioms emerge, are put to the test and combined with one another *inside* mathematical webs (being submitted to various combinatorial, deductive, conceptual and harmonic controls, until they are either discarded or partially accepted).[58] The search for appropriate axioms and criteria of plausibility can thus be seen as self-sufficient, without any need to invoke an external ontology. (A brilliant example of such a methodology is presented in the final chapter of Maddy's book, in studying the axiom of constructibility *V=L* and showing that the axiom internally clashes with basic principles of maximality, ubiquitous in mathematical practice.)

---

57   P. Maddy, *Naturalism in Mathematics* (Oxford: Oxford University Press, 1997).

58   If the explicit term 'mathematics' appears in the titles of both of Maddy's monographs, this is nevertheless restricted to set theory, a fragment of mathematical inquiry.

Gilles Châtelet

Les enjeux du mobile: Mathématique, physique, philosophie [The stakes of the mobile: Mathematics, physics, philosophy][59] directly confronts the fundamental problems of mathematical thought's mobility, and of its *natural osmoses* with physics and philosophy. Châtelet's text puts several sui generis perspectives to work on the spectrum of mathematical philosophy: an opening onto a sort of *primacy of the visual* in mathematical practice (thereby bringing to bear part of Merleau-Ponty's general phenomenological program in the context of mathematics);[60] a special sensitivity to the *mobile emergence* of mathematical concepts and 'things', owing to a study of the *gestures* and processes on the border of the virtual and the actual; careful attention to and subtle analysis of the *metaphorical webs* that accompany the doing of mathematics, and govern its interlacings with physics and philosophy; meticulous study, with detailed concrete cases, of the modes of *articulation* of mathematical knowledge and of

---

59 G. Châtelet, *Les enjeux du mobile: Mathématique, physique, philosophie* (Paris: Seuil, 1993). [Tr. R. Shore & M. Zagha as *Figuring Space: Philosophy, Mathematics and Physics* (Springer, 1999).]

60 On this recuperation of the diagram for the philosophy of mathematics, following the clear French line of filiation – Lautman-Deleuze-Châtelet – see N. Batt, ed., *Penser par le diagramme: De Gilles Deleuze à Gilles Châtelet Théorie-Littérature-Enseignement* 22 (Saint-Denis: Presses Universitaires de Vincennes, 2004) and S. Duffy, ed., *Virtual Mathematics: The Logic of Difference* (Bolton: Clinamen Press, 2006). The latter compilation includes, amid various articles dedicated to logic and mathematics in Deleuze, a posthumous text of Châtelet's (edited by Charles Alunni), 'Interlacing the Singularity, the Diagram and the Metaphor'.

its *dialectical balances*. The titles of the work's five chapters are indicative of Châtelet's originality: 'The Enchantment of the Virtual', 'The Screen, the Spectrum and the Pendulum: Horizons of Acceleration and Deceleration', 'The Force of Ambiguity: Dialectical Balances', 'Grassmann's Capture of the Extension: Geometry and Dialectic', and 'Electromagnetic Space'. Châtelet's array of examples is concentrated in the modern period (Argand, Cauchy, Poisson, Grassmann, Faraday, Maxwell, and Hamilton, among others), but timeless interlacings recur as well (Oresme, De Broglie). In the introduction, Châtelet quotes André Weil's lengthy explanation of the primordial role that 'obscure analogies' play in mathematical investigation – the threshold of creative penumbra that Châtelet explores in approaching the 'gestures that *inaugurate* dynasties of problems', the articulations and torsions between reason and intuition, the 'rational capture of allusions', and the structural and hierarchical deployment of the diagrams of thought. The fourth chapter is something of a gem in the philosophy of mathematics. Châtelet patiently reviews how Grassmann constructs the 'synchronous emergence of the intuitive and the discursive' in a living unity that is neither a priori nor a posteriori, how the dialectic engenders new forms by way of a careful hierarchy of scales in Grassmann's exterior products, how Grassmann's very style leads to a natural approach to the processes that enable the capture of self-reference ('comprehension of comprehension') and how the apparent oppositions continuous/discrete and equal/different consist in *fluxes* of mathematical inventiveness that serve to articulate its various, partial modes of knowing (numbers, combinatorics, functions, extension theory). Going further still, a magisterial thirty-page section on Grassmann products explains, in vivid detail and with the constant presence of diagrams, the great lines of tension of Grassmann's system, which Châtelet explicates in the first part of the chapter. The entire work constitutes a major contribution to the philosophy of mathematics, a contribution to which we will repeatedly return in the third part of this study and that is, to our mind, the most original work on the subject since Lautman's.

Frédéric Patras

La pensée mathématique contemporaine [Contemporary mathematical thought][61] provides an important leap forward in the effort to approach contemporary mathematics. The spectrum traversed is no longer the universe of set theory – which, at the end of the day, is the *customary* spectrum, for all of Badiou's originality and Maddy's expertise – but includes genuinely mathematical aspects (abstract algebra, algebraic geometry, topology, category

---

61   F. Patras, *La pensée mathématique contemporaine* (Paris: PUF, 2001).

theory) and incorporates the rise of modern mathematics (chapters 1–4, with excellent introductions to Galois, to Dedekind's algebra and to the 'universal' Hilbert), as well as aspects of the works of central figures of contemporary mathematics (chapters 5–8, on Bourbaki, Lawvere, Grothendieck, Thom). Chapter 7, dedicated to Grothendieck, is particularly valuable, owing to its sheer singularity among treatises of mathematical philosophy. It should be considered a monumental aberration that a figure who, in all likelihood, is the most important mathematician of the second half of the twentieth century never seems to be seriously considered in 'mathematical philosophy', and Patras seeks to put an end to this error. The author shows that *a comprehension of the modes of emergence of mathematical creativity should constitute one of the indispensable tasks of mathematical philosophy*, and indicates that some of the great forces underlying Grothendieck's work (aesthetic schematization, universal definition, logical cleanliness, inventive 'innocence', 'listening' to the 'voice of things', dialectical yin-yang) can help us understand the mathematical imagination as a form of complex thought, in which multiple structural polarities and bordering tensions interlace.

David Corfield

From its polemical title onwards, *Towards a Philosophy of Real Mathematics* aims to break the normative prejudices that the philosophy of mathematics makes use of,[62] in particular the 'belief amongst philosophers to the effect that the study of recent mainstream mathematics is unnecessary'.[63] A lengthy introduction argues for the value of a philosophical perspective oriented toward nonelementary mathematics, and exhibits some of the major problems that this approach encounters, but that the 'foundationalist filter' still fails to detect: the status of the structural *borders* of mathematics (beyond binarisms and alternatives of the 'all-or-nothing' variety), the *connectivity* of different mathematical theories, the *evolution* of mathematical concepts, the *contingency* of mathematical thought, and the progressive *recursive richness* of mathematical constructions. The subtitle of the introduction, 'A Role for History', indicates the path adopted by Corfield – a junction of mathematics, philosophy and history, in which current reflections on the discipline's development take on a *real* relevance for the philosopher of mathematics. Indeed, the text broaches various themes from contemporary mathematics – automated proofs of

---

62 D. Corfield, *Towards a Philosophy of Real Mathematics* (Cambridge: Cambridge University Press, 2003).

63 Ibid., *Towards a Philosophy of Real Mathematics*, 5.

theorems, modes of indeterminacy, theory of groupoids, *n*-categories – and elaborates an epistemological model in which an intermingling of webs and hierarchies helps to explicate the simultaneously multivalent and unitary development of advanced mathematics. Chapters 2 and 3 deal with logical automata and serve to contrast the limits of automatic proof with groundbreaking mathematical creativity (chapter 4), where the role of analogy turns out to be indispensable for the invention of new concepts, techniques and interpretations (with valuable examples from Riemann, Dedekind, Weil, and Stone). Chapters 5 and 6 review problems of plausibility, uncertainty and probability in mathematics (Bayesian theories) and in science in general (quantum fields). Chapters 9 and 10 approach ongoing developments in mathematics (groupoids, *n*-categories) and the corresponding works of the current investigating mathematicians (Brown, Baez), concretely demonstrating how a mathematics can be observed in utero from a philosophical point of view in which certain traditional ontological and epistemological obstacles have been dissolved. Chapters 7 and 8 focus on the problem of the growth of mathematics (with an appraisal and critique of Lakatos), the importance of opposed mathematical practices *living together*, and the consequent necessity of not discarding from the philosophy of mathematics the supposed *residues* of conceptions of mathematics no longer in vogue.

Corfield tries to make the complex life of mathematics heard (so that we may 'listen to the voice of things', as Grothendieck would write in his *Récoltes et semailles*), beyond which 'one can say with little fear of contradiction that in today's philosophy of mathematics, it is the philosophy that dictates the agenda'. According to Corfield, a healthy inversion of perspectives, to the point where a happy medium can be constructed, could help today's philosophy of mathematics emulate the mental openness of the great Russell by encouraging philosophers to:

1. Believe that our current philosophy is not adequate to make proper sense of contemporary mathematics;
2. Trust that some mathematicians can give us insight into a better philosophical treatment;
3. Believe that the emerging picture will revitalize philosophy.[64]

Some of the examples studied by Corfield indicate how fixing our attention on *more mathematics* (and not necesarily more philosophy, as might narrowly be thought) could help philosophy: the Hopf algebras at the heart of the reasons for mathematics' applicability to quantum physics, the groupoids that display novel interlacings between symmetry (abstract equivalence) and asymmetry

---

64   Ibid., 270.

(noncommutativity), and the categorical languages of Makkai that eliminate poorly posed ontological questions. Altogether, the work supplies an interesting *counterweight* to the dominant forces in the philosophy of mathematics, which are very attentive to language but far removed from 'real' mathematics. The text concludes with an important plea for today's philosophy of mathematics: 'Mathematics has been and remains a superb resource for philosophers. Let's not waste it.'[65]

---

65   Ibid., 270.

2.3 MORE PHILOSOPHY, LESS MATHEMATICS

We have indicated, in sections 2.1 and 2.2, how various philosophers, mathematicians and historians have
approached advanced mathematics (in its three great
realms: classical, modern and contemporary) thereby
opening new perspectives for mathematical philosophy
that have been inexistent or 'effaced' from the point
of view of foundations or of elementary mathematics.
The pretension to exhaust the horizons of mathematical
philosophy with the 'fundamental' and the 'elementary',
and the *unwillingness to see* in modern and contemporary
mathematics an entire arsenal of problematics *irreducible*
to elementary examples or logical discussions (chapter 1),
has limited the reach of the traditional mathematical
philosophy inherited from analytic philosophy. Though
it has neglected the universe of advanced mathematics,
traditional mathematical philosophy has been able to pinpoint complex ontological and epistemological problems
(with respect to the notions of number, set and demonstration), which it has then treated with great precision.

A broad and current vision of traditional mathematical philosophy can be found in *The Oxford Handbook of
Philosophy of Mathematics and Logic*.[66] As we will see in
reviewing the text, the focus is clearly analytical, logical

---

[66] Shapiro, *Oxford Handbook...*

and Anglo-Saxon. Modern and contemporary mathematics, as we have defined these terms, and those who forged modern and contemporary mathematics – Galois, Riemann and Grothendieck, to name only the indispensable figures – make minimal appearances or do not appear at all.[67] By contrast, another of the fundamental figures of modern mathematics, Georg Cantor, is broadly studied throughout the volume, thus underscoring analytic philosophers' interest in set theory. And so the range of *mathematics* reflected upon in the volume is reduced to a lattice of logics and classical set theory. This curious deformation of the mathematical spectrum, which has been repeated for decades now in the Anglo-Saxon world, should no longer be accepted. It would be another matter if, with somewhat more humility, the volume in question had been called *The Oxford Handbook of Analytic Philosophy of Logic*.

Starting with Shapiro's excellent general introduction (where he elaborates on certain remarks from his earlier text, *Thinking about Mathematics*, mentioned in our introduction), the compilation includes a review of the philosophy of mathematics between Descartes and Kant (Shabel); a chapter on empiricism and logical

---

67 The indexes (of both subjects and proper names) at the end of the volume refer only two pages (of the 833 in the volume) to Galois and Riemann; Grothendieck does not even appear. Though the indexes are less than reliable (since, for example, in Steiner's article, which concerns the problem of the applicability of mathematics, Riemann and Galois are studied with greater patience), they are sufficiently indicative of the factual situation.

positivism (Skorupski); an introduction to Wittgenstein's philosophy of logic and 'mathematics' (Floyd); three chapters regarding versions of logicism (Demopoulos and Clark, Hale and Wright, Rayo); one text on formalism (Detlefsen); three chapters on forms of intuitionism (Posy, McCarty, Cook); a text on Quine (Resnik); two chapters on naturalism (Maddy, Weir); two chapters on nominalism (Chihara, Rosen and Burgess); two chapters on structuralism (Hellman, MacBride); one text on the problem of the applicability of mathematics (Steiner), one text on predicativity (Feferman); two chapters on logical consequence, models and constructibility (Shapiro, Prawitz); two chapters on relevance logic (Tennant, Burgess); and two chapters on higher-order logic (Shapiro, Jané). All of the works demonstrate a high level of analysis, extensive argumentative rigor and great professionalism. Nevertheless, what seems to have been created here is an extensive web of cross-references between the authors' professional works and the stratum of logics linked to those works: a secondary web that has been *substituted* for the primary, underlying mathematics. Once this interesting and complex web has been taken up – by means of logical forms, problems associated with foundations, detailed philosophical disquisitions and self-references among specialists – very few of the authors included in the handbook seem sufficiently self-critical to consider that, *perhaps*, many other (*possibly even more interesting and complex*) forms of mathematics have escaped their attention. Of course, we cannot (and should not) ask the specialist to go beyond his field of knowledge, but neither can we (nor should we) confuse the student or professional interested in the topic, fooling him into thinking that the text covers the 'philosophy of mathematics and logic' in its entirety. The *disappearance of mathematics* and its supposed reducibility to logic make up the least fortunate global perspective that Anglo-Saxon analytic philosophy has (consciously or unconsciously) imposed upon the philosophy of mathematics.

It seems surprising that, forty years after the publication of Benacerraf and Putnam's staple anthology *Philosophy of Mathematics*,[68] the problems examined in Shapiro's new compilation remain the same ones treated in the four parts of the 1964 compilation: foundations, mathematical objects, truth, and sets. The tools included in Shapiro's compilation include a much broader and pluralistic web of logics, as well as new unifying perspectives. But the gigantic advances made by mathematics in the last fifty years are dazzling in their absence. Again, it seems as if mathematics has not evolved, as if the problems of the philosophy of mathematics were fixed in time, leaving room only for *scholastic variations*. We hope to show, in

---

68 P. Benacerraf & H. Putnam, eds., *Philosophy of Mathematics: Selected Readings*, 2nd edition (Cambridge: Cambridge University Press, 1984).

the second and third parts of this essay, that the situation we are dealing with here is unsustainable.

With respect to Benacerraf and Putnam's compilation, Shapiro's opens perspectives onto two particular new horizons for mathematical philosophy: naturalism and structuralism. In her article, 'Three Forms of Naturalism', Maddy explores the roots of naturalism in Quine, and the later modifications of Quinean positions in Burgess and in Maddy's own work.[69] Quine's self-referential naturalist position, according to which the foundations of a science and its fragments of certainty should be sought in the science itself, and not in a first philosophy that is external and alien to the science, provokes a robust *intramathematical* perspective in Maddy, according to which a naturalist philosopher of mathematics should not slide into extramathematical metaphysical debates, but must meticulously track the dynamics of concept formation *within* her own discipline. Maddy has satisfied this program with vigor and originality *within set theory*, showing, in particular, that the supposedly Quinean naturalist position in favor of a reduced universe of sets ($V=L$) receives no sympathy from the 'natural' arguments in favor of large cardinals, conducted by the theory's chief creators (Martin, Woodin and Shelah, among others). Nevertheless, the 'mathematics' that the philosopher deals

---

69   Shapiro, *Oxford Handbook...*, 437–59.

with here is restricted, once again, to forms of logic and set theory, without making any inroads into geometrical, algebraic or differential domains, and without coming close to mentioning any of the Fields medalists (except for Cohen, of course) who, presumably, have changed the course of the discipline over the last fifty years.

In his 'Structuralism' article, Geoffrey Hellman proposes four versions of a structural focus on mathematics (set-theoretical structuralism, generic structuralism, categorical structuralism and modal structuralism), and goes on to compare the advantages of each version with respect to certain philosophical problems that arise within the structural aspects themselves. These problems include the following: the contrast between 'set' and 'structure' and the choice of natural axiom-concepts; 2. the handling of 'totalities'; the emergence of intractable 'ontologies-epistemologies'; the handling of rigid and nonrigid structures from a philosophical perspective; the presence of circularities in structures; the problems of theoretical under-determination; the presence of primitive, undefined conceptual substrata.[70] Hellman's conclusions (carefully delimited, in the style of all the authors of the handbook, with sound lines of argumentation and with reference to a minimum of mathematical cases) indicate that a *mixture of categorical and modal structuralism* could respond in the

---

70 Ibid., 536–62.

best way possible to the problems confronted in the article. We shall see, in part 3 of this book, how to construct and significantly *extend* that mixture, which is suggested by Hellman and *reclaimed* by the extensive case studies that we will undertake in the second part.

The analytic school of philosophy of mathematics, including, in particular, the great majority of its Anglo-Saxon practitioners (with important exceptions, of course[71]), could feel at home under the slogan, 'more philosophy, less mathematics'. This has always been a perfectly valid option, but a restrictive one as well, no doubt. The *danger* – which has always existed, and continues to exist, and which Rota emphatically opposed – is that in many academic environments, this option is the *only one* available. Returning to behold, again, the complexity of the mathematical world – as Lautman admirably succeeded in doing, along with many of the authors reviewed in section 2.2, and which Corfield has again proposed as an *imperative* – should reset the balance, and put forward a new plan of greater equality: '*as much* mathematics *as* philosophy.' Part 2 of this work aims to cover the left side of the balance; part 3, the right.

---

71   In addition to the authors mentioned in section 2.2, we could point out other Anglo-Saxon philosophers and historians who try to cover a broad mathematical spectrum (methodological, technical and creative), such as Jeremy Gray, Michael Hallett, Mark Steiner and Jamie Tappenden, among others.

CHAPTER 3

TOWARD A SYNTHETIC PHILOSOPHY OF CONTEMPORARY MATHEMATICS

In the introduction and preceding chapters we saw that a contrasting (and often contradictory) *multiplicity* of points of view traverse the field of the philosophy of mathematics. Also, we have delineated (as a first approximation, which we will go on to refine throughout this work) at least five characteristics that separate modern mathematics from classical mathematics, and another five characteristics that distinguish contemporary mathematics from modern mathematics. In that attempt at a global conceptualization of certain mathematical tendencies of well-defined historical epochs, the immense *variety* of the technical spectrum that had to be traversed was evident. Nevertheless, various reductionisms have sought to limit both the philosophical multiplicity and mathematical variety at stake. Far from *one* kind of omnivorous philosophical wager, or *one* given reorganization of mathematics, which we would then try to bring into a univalent correlation, we seem to be fundamentally obliged to consider the necessity of constructing *multivalent* correspondences between philosophy and mathematics, or rather between philosophies and mathematics in the plural.

In a manner consistent with this situation, we will not assume any a priori philosophical position until we have carefully observed the contemporary mathematical landscape. We will, however, adopt a precise methodological framework, which, we believe, will help us better observe that landscape. Of course, that methodological schematization will also influence our modes of knowing, but we trust that the distortions can be controlled, since the method of observation we will adopt and the spectrum we presume to observe are sufficiently close to one another. The philosophical and mathematical consciousness of multiplicity at stake, in fact, requires a minimal instrumentarium that is particularly sensitive to the transit of the multiple, that can adequately take stock of that multiplicity, and that allows us to understand its processes of translation and transformation. To those ends, we will adopt certain minimal epistemological guidelines, furnished in philosophy by Peirce's pragmatism, and in mathematics, by category theory.

A vision moderately congruent with the multiformity of the world should integrate at least three orders of approximations: a *diagrammatic* level (schematic and reticular) where the skeletons of the many correlations between phenomena are sketched out; a *modal* level (gradual and mixed) where the relational skeletons acquire the various 'hues' of time, place and interpretation; and a *frontier* level (continuous) where webs and mixtures are progressively combined. In this 'architecture' of vision, the levels are never fixed or completely determined; various contextual saturations (in Lautman's sense) articulate themselves here (since something mixed and saturated on a given level may be seen as skeletal and in the process of saturation in another, more complex context) and a dynamic frontier of knowledge reflects the undulating frontier of the world. An adequate integration of diagrams, correlations, modalities, contexts and frontiers between the world and its various interpretants is the primordial object of *pragmatics*. Far from being the mere study of utilitarian correlations in practical contexts of action-reaction (a degeneration of the term 'pragmatics' that corresponds to the disparaging way in which it gets used these days), pragmatics aims to reintegrate the differential fibers of the world, explicitly inserting the broad relational and modal spectrum of fibers *into* the investigation as a whole. The technical attention to contextualizations, modulations and frontiers affords pragmatics – in the sense which Peirce, its founder, gave it – a fine and peculiar methodological timbre. Just as vision, like music, benefits from an integral modulation through which one interlaces tones and tonalities so as to create a texture, so pragmatics benefits from an attentive examination of the contaminations and osmoses between categories and frontiers of knowledge so as to articulate the diversity coherently.

Various natural obstructions are encountered on the way to any architectonic system of vision seeking to reintegrate the Many and the One without losing the multivalent richness of the differential. One obvious obstruction is the impossibility of such a system's being stable and definitive, since no given perspective can capture all the rest. For, from a logical point of view, whenever a system observes itself (a necessary operation if it seeks to capture the 'whole' that includes it), it unleashes a self-referential dynamic that ceaselessly hierarchizes the universe. As such, a pragmatic architectonic of vision can only be *asymptotic*, in a very specific sense interlacing evolution, approximation and convergence, but without requiring a possibly nonexistent limit. An 'internal' accumulation of neighborhoods can indicate an orientation without having to invoke an 'external' entity that would represent a supposed 'end point' – it has *the power to orient ourselves within the relative without needing to have recourse to the absolute*. This fact harbors enormous consequences, whose full creative and pedagogical force is just beginning to be appreciated in the contemporary world.

The maxim of pragmatism – or 'pragmaticism' (a name 'ugly enough to escape the plagiarists'[72]) as Peirce would later name it in order to distinguish it from other behaviorist, utilitarian and psychologistic interpretations

---

72 C. S. Peirce, *Collected Papers* (Harvard: Harvard University Press, 1931–1958), vol. 5, 415.

– appears to have been formulated several times through-out the intellectual development of the multifaceted North American sage. The statement usually cited is from 1878, but other more precise statements appear in 1903 and 1905:

Consider what effects that might conceivably have practical bearings, we conceive the object of our conception to have. Then our conception of those effects is the whole of our conception of the object. (1878)[73]

Pragmatism is the principle that every theoretical judgment expressible in a sentence in the indicative mood is a confused form of thought whose only mean-ing, if it has any, lies in its tendency to enforce a corre-sponding practical maxim expressible as a conditional sentence having its apodosis in the imperative mood. (1903)[74]

The entire intellectual purport of any symbol consists in the total of all general modes of ratio-nal conduct, which, conditionally upon all the possible different circumstances and desires,

---

73 Peirce, 'How to Make Our Ideas Clear' (1878), in *Collected Papers*, vol. 5, 402.

74 Peirce, 'Harvard Lectures on Pragmatism' (1903), in *Collected Papers*, vol. 5, 18.

would ensue upon the acceptance of the symbol. (1905)[75]

What is emphasized in the 1905 statement is that we come to know symbols according to certain 'general modes', and by traversing a spectrum of 'different possible circumstances'. This modalization of the maxim (underscored in the awkward repetition of 'conception' in 1878) introduces into the Peircean system the problematic of the 'interlacings' between the *possible* contexts of interpretation that may obtain for a given symbol. In the 1903 statement, on the one hand, we see that every practical maxim should be able to be expressed in the form of a conditional whose *necessary* consequent should be adequately contrastable, and on the other hand, that any indicative theoretical judgment, in the *actual*, can be specified only through a series of diverse practices associated with that judgment.

Expanding these precepts to the general field of semiotics, to know a given sign (the realm of the *actual*) we must traverse the multiple contexts of interpretation capable of interpreting that sign (the realm of the *possible*) and, in each context, study the practical imperative consequences associated with each one of those interpretations (the realm of the *necessary*). Within that general landscape, the *incessant and concrete transit* between the

---

75   Peirce, 'Issues of Pragmaticism' (1905), in *Collected Papers*, vol. 5, 438.

possible, the actual and the necessary turns out to be one of the *specificities* of mathematical thought, as we will repeatedly underscore throughout this work. In that transit, the relations between possible contexts (situated in a *global* space) and the relations between the fragments of necessary contradistinction (situated in a *local* space) take on a primordial relevance – something that, of course, finds itself in perfect tune with the conceptual importance of the logic of relations that Peirce himself systematized. In this way, the pragmaticist maxim indicates that knowledge, seen as a logico-semiotic process, is preeminently contextual (versus absolute), relational (versus substantial), modal (versus determined), and synthetic (versus analytic).

The maxim filters the world through three complex webs that allow us to differentiate the One into the Many and, inversely, integrate the Many into the One: the aforementioned *modal* web, a *representational* web and a *relational* web. In effect, besides opening onto the world of possibilities, the signs of the world should, above all, be representable in the languages (linguistic or diagrammatic) utilized by communities of interpretants. The problems of representation (fidelity, distance, reflexivity, partiality, etc.) are thus immediately bound up with the *differentiation of the One and the Many*: the reading of an identical fact, or an identical concept, dispersed through many languages, through many 'general modes' of utilizing information, and through many rules for the organization and stratification of information.

One of the strengths of Peircean pragmatism, and in particular of the fully modalized pragmaticist maxim, is that it allows us to *once again reintegrate the Multiple into the One* through the third web that it puts in play: the relational web. In fact, after decomposing a sign into subfragments in the various possible contexts of interpretation, the correlations between fragments give rise to new forms of knowledge that were buried in the first perception of the sign. The pragmatic dimension emphasizes the *coalition* of some possible correlations, discovering analogies and transfusions between structural strata that, prior to effecting that differentiation, had not been discovered. In this way, though the maxim detects the fundamental importance of local interpretations, it also insists on the reconstruction of global approximations by means of adequate *gluings* of the local. We shall later see how the tools of the mathematical theory of categories endow these first vague and general ideas with great technical precision. The pragmaticist maxim will then emerge as a sort of abstract *differential and integral calculus*, which we will be able to apply to the general theory of representation, that is to say – to logic and semiotics in these sciences' most generic sense, the sense foreseen by Peirce.

Overleaf we present a diagrammatic schematization of the pragmaticist maxim, in which we synthetically condense the preceding remarks. This diagram (*figure 4*) will be indispensable for *naturally* capturing the maxim's structuration from the perspective of mathematical category theory. Reading from left to right, the diagram displays an actual sign, multiply represented (that is, underdetermined) in possible contexts of interpretation, and whose necessary actions-reactions in each context yield its partial comprehensions. The terms 'pragmatic differentials' and 'modulations' evoke the first process of differentiation; the latter term reminds us of how a *single* motif can be extensively altered over the course of a musical composition's development. The process of reintegration proper to Peircean pragmatics is evoked by the terms 'pragmatic integral' and 'correlations', 'gluings', 'transferences', which remind us of the desire to return that which has been fragmented to a state of unity. The pragmatic dimension seeks the *coalition* of all possible contexts and the integration of all the differential modulations obtaining in each context, a synthetic effort that has constituted the fundamental task of model theory and category theory in contemporary logic.

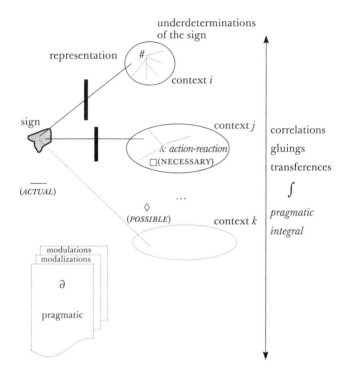

Figure 4. *Sketch of Peirce's pragmaticist maxim.*

The pragmaticist maxim thus serves as a sophisticated 'sheaf of filters' for the decantation of reality. The crucial role of the sheaf secures an amplified multiplicity of perspectives which, for that matter, filters information in more ways than one, thus establishing *from the outset* a certain *plausibility* for the claim that knowledge may be sufficiently rich and multivalent. The Peircean pragmaticist maxim may come to play an extraordinarily useful role in the philosophy of contemporary mathematics. Its first upshot consists in *not privileging* any point of view or any fragment of language over another, thereby *opening* the possibility of considering what Susan Haack called 'rival, incompatible truths', without supposing them to be 'reducible to a privileged class of truths in a privileged vocabulary'.[76] In a pendular fashion, the second crucial strength of the maxim consists in the power to *compare* very diverse *levels* within the multiplicity of perspectives, languages and contexts of truth that it has succeeded in opening up. Indeed, the unwillingness to restrictively assume any privileged 'foundation' does not oblige us to adopt an extreme relativism without any hierarchies of value. In questions of foundations, for example, not reducing things to discussions referring to the supposedly 'absolute' base ZF or the dominant force of first-order classical logic, but opening them instead to discussions in other deductive *fragments* ('reverse mathematics'), semantic *fragments* (abstract model theory) or structural *fragments* (category theory) of ZF, is a strategy that broadens the contexts of contradistinction – and, therefore,

---

76 S. Haack, C. Bo, 'The Intellectual Journey of an Eminent Logician-Philosopher', in C. de Waal, ed., *Susan Haack: A Lady of Distinctions* (Amherst, NY: Prometheus Books, 2007), 27.

the *mathematical* richness at stake – without, for all that, spilling into epistemological disorder. Our contention is just the contrary: it is *in virtue of* being able to escape various 'ultimate' foundations, and *in virtue of* situating ourselves in *a relative fabric of contradistinctions*, *obstructions*, *residues and gluings*, that a genuine epistemological *order* for mathematics asymptotically arises and evolves.

The Peircean pragmaticist maxim can be seen as a sophisticated way of weaving between analysis/differentiation and synthesis/integration. The contemporary world requires new conceptual instruments of 'collation' or 'gluing' (which respond with new arguments to the 'primordial' differentiation/integration dialectic[77]), and, to a large extent, the Peircean pragmaticist maxim provides one of those gluing instruments. As we understand it, if there is any mathematical concept capable of serving as a threshold between modern and contemporary mathematics, it is that of a *mathematical sheaf*, which is indispensable for reintegrating adequate local compatibilities into a global gluing.[78] Correlatively, within the scope of epistemology, we believe that the Peircean pragmaticist

---

77 A good presentation of the analysis/synthesis polarity and its *subsumption* into the 'greater' differentiation/integration polarity, can be found in Gerald Holton's article, 'Analisi/sintesi', for the *Enciclopedia Einaudi* (Torino: Einaudi, 1977), vol. 1, 49–522.

78 We will study, in detail, the multiple facets of sheaves in topology, algebraic geometry and logic in the second part of this work. The mobile plasticity of sheaves not only lets us pass from the local to the global, but, in a natural fashion, allows for multiple osmoses between very diverse subfields of mathematics. In a certain way, since their very genesis, sheaves have acquired an incisive *reflexive richness* that has rendered them extraordinarily malleable.

maxim can serve as a remarkable *methodological sheaf* to compare and then interlace the diverse. The Peircean web of webs, in effect, opens onto the modal realms in their *entirety*, and systematically attends to contrasting given facts (within the phenomenological world) and necessary behaviors (within well-defined contextual systems), so as to then reintegrate them in an extended spectrum of possible signs.

Differentiation and reintegration reach a high degree of methodological precision in the mathematical theory of categories. As a counterpart to the set-theoretical analytic championed by Cantor's heirs,[79] category theory no longer dissects objects from within and analyses them in terms of their elements, but goes on to elaborate synthetic approaches by which objects are studied through their external behavior, in correlation with their ambient milieu. Categorical objects *cease to be* treated analytically and are conceived as 'black boxes' (with local Zermelo-Von Neumann elements becoming invisible, while global elements emerge). Their movement through variable contexts is observed by means of the significant accumulative effort of synthetic characterizations. Within certain classes of structures (logical, algebraic, ordered, topological, differentiable, etc.), category theory detects general synthetic

---

79 Cantor himself is better situated in terms of a sort of general *organicism* (with considerable and surprising hopes that his *alephs* would help us understand both the living realm and the world around us), where analytic and synthetic considerations relate to one another.

invariants and defines them by means of certain 'universal properties'. Those properties, in the first instance, hold for given universes of classes of structures (= 'concrete' categories), but can often be extended to more general fields in which the minimal generic properties of those classes are axiomatized (= 'abstract' categories). Between abstract and concrete categories, multiple weaves of information (= 'functors') are then established. An incessant process of *differentiation* diversifies the universal constructions given in abstract categories and, in contrasting forms, 'incarnates' them in multiple concrete categories. Inversely – in a pendular fashion, we might say – an incessant process of *integration* seeks out common constructors and roots, at the level of abstract categories, for a great variety of the special constructions showing up in concrete categories.

In this way, a *quadruple* synthetic strategy takes shape in category theory. First of all, internally, in each concrete category, we seek to characterize certain special constructions in terms of their environmental properties in the given class. Then, externally, in the general field of abstract categories, we seek out certain universal constructions that can account for the characterizations obtained in the concrete categories. In the third stage, in a remarkable *weaving* between concrete and abstract categories, we go on to define adequate functors of differentiation and reintegration. Finally, the same functors become the object of investigation from a synthetic point of view, and their osmoses and obstructions (= 'natural transformations') are studied systematically. Category theory, as we will see in chapters 4–7, has acquired considerable mathematical value in its own right, but for the moment we are interested only in accentuating its methodological interest for a philosophy of mathematics *open* to incessant pendular processes of differentiation and reintegration.

Indeed, *if the philosophy of mathematics could make use of the synthetic lessons on differentiation and reintegration codified in both the Peircean pragmaticist maxim and in the functorial processes of category theory*, many of the fundamental problems in philosophy of mathematics might acquire new glints and twists that, we believe, could enrich philosophical dialogue. The objective of this essay's third part will be precisely to discuss those problems, in light of the contributions of contemporary mathematics, and in light of a synthetic grafting of the Peircean pragmaticist maxim onto the methodological lineaments of category theory. However, in posing the *same* problematics from the complementary perspectives of analysis and synthesis, we can already indicate certain fundamental *inversions* (see figure 5, overleaf) in the *demands* forced upon us by analytic and synthetic perspectives.

PROBLEMATIC	(PHILOSOPHY OF LANGUAGE + SET-THEORETICAL FOUNDATIONS)	CATEGORY-THEORETIC CONTEXTS)
realist ontology: mathematical objects exist in a real world	must postulate the real existence of the universe of sets, to which we are granted access by a reliable form of intuition	must postulate the existence of a covering of the real by means of progressive hierarchies of structural contexts that asymptotically approximate it
idealist ontology: mathematical objects are linguistic subterfuges	must postulate a dissociation between mathematical constructs and their physical environments	must postulate a dissociation between classes of linguistic categories and classes of categories from mathematical physics
realist epistemology: truth values reflect objective forms of knowledge	must postulate the existence of a set-theoretic semantics as an adequate transposition of semantic correlations in the real world	must postulate the existence of categorical semantic adjunctions and invariant skeletons persisting through functorial weavings
idealist epistemology: truth values are subjective forms of control	must postulate a variability of modalization of sets, and assume the existence of stable transitions between 'compossible' worlds	must postulate the impossibility of 'archetypical' initial categories capable of generically classifying the truths of their derived categories
realist metaphysics: to ti en einai ('the essential of essence') exists mathematically	must postulate the existence of a 'monstrous' model and reflexion schemas that would accommodate every universe of sets	must postulate the existence of multiple classifier toposes and additional inverse classifier where the classifiers can be 'glued together'
idealist metaphysics: to ti en einai does not exist mathematically	must postulate the necessity of towers of set-theoretical universes that can be controlled only through relative consistencies	must postulate the necessity of functorial iterations ad infinitum, irreducible to projections from a supposedly 'final' classifier

Figure 5 (facing page). *Complementary perspectives on the 'pure' setting out of problematics in the philosophy of mathematics.*

As we shall see and discuss in part 3, some of the above requirements seem too strong and *go against the grain* of various advances achieved in contemporary mathematics. For example, from a synthetic point of view – which is better suited than an analytic one to mathematical practice – an idealist ontology that dissociates linguistic categories (à la Lambek) from categories of mathematical physics (à la Lawvere) seems *inviable*, because it places itself in immediate contradiction with advances in *n*-categories (à la Baez) that allow us *simultaneously* to account for complex torsors in linguistics and physics. Another example, again with a synthetic focus, seems to show that an idealist epistemology is similarly *inviable*, since it would conflict with the (already actualized) possibility of constructing classifier topoi and initial allegories (à la Freyd). In this manner, it is thus easy to see how our alternative, double strategy – *to make use of 'synthetic' methodological foci and work closely with contemporary mathematics* – can bear considerable philosophical fruit. We hope to show, further on, that directing our attention to *more mathematics* (and not necessarily 'more philosophy') represents a reasonable strategy, and one that opens attractive and unexpected channels for philosophical dialogue.

The Peircean pragmaticist maxim and the methodological lineaments of category theory help to provide a vision of *mathematical practice* that is *fuller and more faithful* than what an analytic vision offers. The reasons are varied and have to do with the meanings ordinarily given to the terms 'full' and 'faithful' – provided we extend their scope in the direction of the meanings they take on in the technical context of category-theoretic functors. Observing that every construction that is realized in a given mathematical environment (topological, algebraic, geometrical, differential, logical, etc.) is necessarily local in an adequate context,[80] we will call a context in which the construction can be locally realized, but which does not, in addition, invoke redundant global axioms, a *minimal context of adequation*. When a vision of a determinate mathematical environment allows us to associate a minimal context of adequation with every mathematical construction in the environment, we will say that the vision is *full*. We will say that a vision of a mathematical environment is *faithful* when it allows us to reconstruct every mathematical construction of the environment in a minimal context of adequation. The *fullness* of the vision ensures that the local richness of the theories will not be diluted in a global magma; the *faithfulness* of

---

80 The context of adequation can be very large: If the mathematical construction is, for example, the cumulative universe of sets, a context rendering the cumulative hierarchy local will have to reach some inaccessible cardinal or other. Nevertheless, the majority of 'real' mathematical constructions (in Hardy's sense, taken up again by Corfield) live in mathematical contexts that are under far greater control, with respect to both cardinal and structural requirements.

the vision ensures that the local richness is really sufficient for its full development. For example, the usual analytic vision of mathematics – based on ZF set theory and its underlying first-order classical logic – turns out to be *neither full nor faithful* in this sense. Given the broad global reach of the ZF axiomatic, the vision is not full, precisely because the minimal contexts of adequation are forcibly lost (a situation of *information loss* – that is to say, a loss of fullness – for which reverse mathematics proposes a palliative); nor is it faithful, since most of the constructions are realized by means of an uncontrolled invocation of the full force of the axioms.

By contrast, mathematical practice turns out to be much closer to a vision that genuinely and persistently seeks to detect, between minimal contexts of adequation, *both transferences and obstructions alike*. The notions of obstruction and residue are fundamental here, since the incessant survey of obstructions, and the reconstruction of entire maps of mathematics on the basis of certain residues attached to those obstructions, is part and parcel of both mathematical inventiveness and its subsequent demonstrative regulation.[81] Now, the obstructions and

---

81 Riemann's $\zeta(s)$ function provides an exemplary case, here. From its very definition (by analytic extension, surrounding its singularities in the line $Re(s)=1$), to its still mysterious applicability in number theory (clustered around the proof that the zeros of the Z-function lie on the line $Re(s)=1/2$), the ζ *extends* its domains of invention and proof *in virtue of the obstructions* – as much definitional as structural – on which mixed constructions of great draught are dashed (here, the ζ function as a 'hinge' between number theory, complex analysis and algebraic geometry).

residues acquire meaning only locally, with respect to certain contexts of adequation – something of which the usual analytic vision often loses sight, and of which, by contrast, a synthetic vision helps us take stock. As we saw in the 'map' of the Peircean pragmaticist maxim (figure 4, p. 118), we are also dealing with a situation that is particularly susceptible to being detected by the maxim, insofar as the latter attends to local differentials and contextual singularities, no less than to the subsequent modal reintegration of local fractures.

A *synthetic philosophy of contemporary mathematics* must therefore seek to capture at least the following *minimal* characteristics that naturally arise in a sort of *generic* 'differential and integral methodology', in which mathematics, philosophy and history are interlaced:

1. a contextual and relational delimitation of the field of contemporary mathematics with respect to the fields of modern and classical mathematics;

2. a differentiation of the plural interlacings between mathematics and philosophy, followed by a reintegration of those distinctions in partial, unitary perspectives;

3. a presentation of a full and faithful vision of mathematical practice, particularly sensitive to a pendular

weaving between transferences and obstructions, and between smoothings and residues;

4. a diagramming of the multivalences, ramifications and twistings between spectra of mathematical theorems and spectra of philosophical interpretations.

In what follows, we will take up precise case studies in contemporary mathematics, by means of which we will be able to repeatedly emphasize these four points, before returning, in part 3, to additional 'skeletal'[82] considerations concerning the philosophy of mathematics.

---

82 In our strategy, one can observe an approximate analog to the practice in category theory whereby, firstly, a category is delimited from other neighboring categories, secondly, various concrete constructions of the category are studied in detail, and thirdly, its skeleton and its free constructions are finally characterized. The three parts of our work correspond – by an analogy that is not overly stretched – to the study of the 'category' of contemporary mathematics and its various adjunctions with respect to the various 'categories' of philosophical interpretations.

PART TWO

Case Studies

CHAPTER 4

GROTHENDIECK: FORMS OF HIGH MATHEMATICAL CREATIVITY

In this second part, we present brief case studies from the landscape of contemporary mathematics (1950–2000). Our strategy will consist in providing mathematical information (primarily conceptual and, to a lesser extent, technical information) that is usually taken to fall outside the scope of philosophy – information whose philosophical distillation and discussion will occupy us in part 3. Nevertheless, although the primary objective of this second part is to expand the *concrete* mathematical culture of the reader, we will also go on to indicate and briefly discuss a few *generic* lines of tension, both methodological and creative, that a complete philosophical understanding of mathematics will have to confront. Contemporary mathematics has given rise to new forms of transit in knowledge, which, in turn, generate new philosophical problems, and new partial solutions of those problems.

4.1 GROTHENDIECK'S LIFE AND WORK: A BROAD OUTLINE

Alexander Grothendieck was born in Berlin (1928), where, during his early childhood (1933–39), he is educated under the care of a Lutheran minister (Heydorn), while his parents actively dedicate themselves to political agitation. His father, Alexander Shapiro, Russian anarchist, radical in Germany and France during the twenties and thirties, Brigadista in the Spanish Civil War, is murdered in Auschwitz. His mother, Hanka Grothendieck – journalist for left-wing magazines, Shapiro's comrade in France and Spain – is reunited with her son after the defeat of the Spanish Republic.[83] Between 1940 and 1942, Alexander and his mother are interned in the Rieucros Concentration Camp, which he will later able to leave for Le Chambon in order to be placed under the care of another protestant pastor (Trocmé) until the end of the war. Reunited with his mother, Alexander completes his degree in mathematics at the University of Montpellier, where several of his teachers remark on his 'extraordinary ability, unsettled by suffering'. It is then that the brilliant young man, unhappy with the calculus being taught to him at the university, proposes a complete theory of

---

83 A good overview of Grothendieck's life can be found in A. Jackson, '*Comme Appelé du Néant* – As if Summoned from the Void: The Life of Alexander Grothendieck', *Notices of the AMS* 51 No. 4, 10, 2004: 1037–56, 1196–1212. A forthcoming biography of Grothendieck, by Colin McLarty, should begin to fill in an inexcusable lacuna.

integration which, although he doesn't know it, turns out to be equivalent to Lebesgue's theory.

From that day forward, it is the incessant *making* of mathematics, rather than its study, that occupies Grothendieck.[84] He is initiated into higher mathematics while participating (in 1948) in the Cartan Seminar at the Ecole Normale Supérieure, completes his doctoral thesis[85] under Dieudonné and Schwartz at Nancy between 1949 and 1953, and then visits America (São Paolo, 1953–54; Kansas, 1955), where he becomes a well-known specialist in topological vector spaces.[86] He then invents K-theory in 1957 and proposes a profound generalization of the Riemann-Roch Theorem, with important consequences for the mathematics of the late fifties and early sixties. (We will expand on K-theory and the Riemann-Roch-Grothendieck

---

84 There is a famous anecdote about a visitor who had been to the Institut des Hautes Études Scientifiques (IHES), (created for Grothendieck in the sixties) and had been struck by the poverty of the library at such a Mecca of mathematics. Grothendieck answered him, 'We don't read mathematics, here; we make mathematics.'

85 According to Dieudonné – an expert on analysis, if there ever was one – Grothendieck's thesis could only be compared, in the field of topological vector spaces, with the works of Banach.

86 *Nuclear spaces*, introduced by Grothendieck in his doctoral thesis, are topological vector spaces defined by families of seminorms with a telescopic property (every unit ball of a seminorm can be embedded in the balls of the remaining seminorms by means of adequate multiplications). What is at stake here are spaces that capture, in a natural way, important families of functions in complex and differential analysis (entire holomorphic functions, smooth functions over compact differential varieties), and that trace, in the infinite, certain good properties of finite-dimensional spaces. The treatments of those properties by means of tensorial products begins to concretize some of Grothendieck's later grand strategies: to study the properties of an object by inserting it in a class (*category*) of similar objects; to construct *transmitters* of information for the properties of an object; to *compare* similar behaviors in other categories, and reutilize all of the *pendular* information accumulated in order to *capture* the initial object in a new light.

theorem in chapter 6, as we approach Atiyah's work.) In 1957 he also publishes his treatise-article, 'On a Few Points of Homological Algebra' (on which we shall comment in section 4.3), where he presents his program for the renovation of algebraic geometry.[87]

In the sixties, the IHES, with Grothendieck at the helm, becomes the world's leading center for mathematical inquiry. What ensues is a decade of creation, on the basis of his central, driving ideas – schemes, topoi, motifs – with the production of the two great series of writings that would completely renovate the mathematics of the age: the *Elements of Algebraic Geometry* (EAG)[88] and the *Seminar on Algebraic Geometry* (SAG).[89] Grothendieck receives the Fields Medal in 1966, and in the panorama of subsequent medalists, his spectrum of influence is enormous (figure 6). Although he surprisingly retires from the mathematical world in 1970 (at 42 years of age!), after having left behind a body of work that whole schools of mathematicians would be hard-pressed to produce in a century, he goes on to write great mathematical manuscripts[90] and

---

87 A. Grothendieck, 'Sur quelques points d'algèbre homologique', *Tôhoku Math. Journal* 9, 1957: 119–221. The article is usually known as 'Tohoku', after the periodical in which it was published.

88 A. Grothendieck (edited in collaboration with J. Dieudonné), *Éléments de Géométrie Algébrique*, 4 vols., 8 parts (Paris: IHES, 1960–67).

89 A. Grothendieck et al., *Séminaire de Géométrie Algébrique du Bois-Marie*, 7 vols., 12 parts (Berlin: Springer, 1970–3), original mimeographed fascicles, 1960–9.

90 *La longue marche à travers la Théorie de Galois* [The Long March through Galois Theory], 1981, 1,800 pages. *Esquisse d'un programme* [Sketch for a Program], 1983, a sort of mathematical testament, 50 pages. *Les dérivateurs* [Derivators], 1990, 2,000 pages.

interminable (self-)critical reflections[91] on the world (both mathematical and theological). In sum, the body of work left behind by Grothendieck is gigantic, both in terms of its depth (the mathematics of the period 1970–2000, particularly the Fields panorama, can to a large extent be seen as a sort of 'commentary' on Grothendieck) and its quantity (about ten thousand manuscript pages). What we are dealing with here is a genuine gold mine for commentators and philosophers, who have barely begun to approach it.[92]

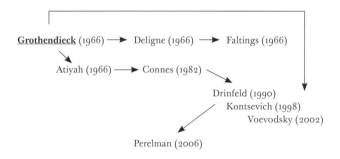

Figure 6. *Grothendieck's lines of influence in the panorama of Fields medalists.*

---

91 *Récoltes et Semailles* [Reaping and Sowing], 1985–6, 1,000 pages. *La clef des songes* [The Key to Dreams], 315 pages.

92 Several digitized fragments of Grothendieck's work, accompanied by a few studies, can be found on the website http://www.grothendieckcircle.org, maintained by Leila Schneps and Pierre Lochak.

After the remarkable fifties (nuclear spaces, K-theory, homological algebra), Grothendieck's first great driving idea during the golden age of the IHES would propel a profound renovation of algebraic geometry. Situating himself within what Thom would later call the 'founding aporia of mathematics'[93] (that is to say, within the *irresolvably contradictory dialectic*, discrete/continuous), Grothendieck invents his *schemes* as a very powerful tool in an attempt to resolve the Weil Conjectures (1949). A precise grafting of the discrete and the continuous, the conjectures seek out a way to measure the number of points in certain algebraic varieties over finite fields, by means of certain generating functions, such as the zeta functions originating in Riemann's continuous, complex-topological intuition.[94] Dwork (1960) demonstrated the rationality of the zeta functions, Grothendieck (1966) the functional equation that governs them, and Deligne (1974), Grothendieck's greatest student, the adequate distribution of their zeros (which gives us combinatorial control of the points in a variety). Deligne's result, which won him the Fields Medal, is a genuine technical *tour de force*.

Modern mathematics, in the first half of the twentieth century, culminates in Weil's astonishing exploration.

---

93 R. Thom, 'L'aporia fondatrice delle matematiche', *Enciclopedia Einaudi* (Torino: Einaudi, 1982), 1133–46.

94 A. Weil, 'Numbers of solutions of equations in finite fields', *Bul. Am. Math. Soc.* 55, 1949: 497–508.

Driven by a very subtle and concrete intuition, as well as an unusual capacity for uncovering analogies at the crossroads between algebraic varieties and topology, Weil succeeded in *stating* his conjectures with great precision. Contemporary mathematics, in the second half of the twentieth century, emerges with the work of Grothendieck, giving rise to the entire apparatus of algebraic geometry that will allow it, in turn, to *resolve* those conjectures. While Zariski's topologies serve as mediations at the algebraic varieties/topologies crossroads, and allow the conjectures to be stated, the ('étale', *l*-adic) cohomologies of Grothendieck and his school serve as mediations at the schemes/topos crossroads, by means of which they can now be resolved. Extending the algebraic varieties into the field of schemes, the richness of Grothendieck's *generic invention* is not the outcome of gratuitous generalizations. At no point is generalization carried out without adequate particularizations in mind, and what is really at stake is a complex process of *ascent and descent* that, as we shall see in greater detail in section 4.2, turns out to be constantly governed by concrete consequences of ever greater mathematical significance.

Indeed, with the creation of his schemes, Grothendieck grafts together two of the major currents of modern mathematics: *the vision of Riemann*, which allows us to understand a curve X by means of the *ring $M(X)$* of the meromorphic functions over that curve, and *the vision* *of Galois-Dedekind*, which allows us to understand an algebraic variety V by means of the spectrum $Spec(V)$ of its maximal ideals. In effect, Grothendieck generalizes the situation so as to be able to *envelope* both visions at once, and suggests that we might understand an *arbitrary* (commutative, unitary) ring by means of a hierarchy of three extremely mathematically rich objects: the spectrum of its prime ideals, the Zariski topology over the spectrum of primes, and the natural sheaves over the topologized spectrum. Those sheaves, with certain additional conditions on their fibers, turn out to be the 'schemes' (*schémas*) of Grothendieck, who thus succeeds not only in unifying some of modern mathematics' deepest intuitions (Galois, Riemann), but in doing so, *broadens the very conception of space*, such that we are no longer concerned with points, but with positions and movement (sections of the sheaf).

From its broad, general outlines to its most particular technical concretizations (as we shall see in section 4.3), Grothendieck's work yields a fundamental paradigm, which we should like to call the *practice of a relative mathematics*. Grothendieck's strategies can indeed be understood, in a conceptual sense, as close to the relativistic modulations that Einstein introduced into physics. In a technical manner, both Einstein and Grothendieck manipulate the frame of the observer and the partial dynamics of the agent in knowledge. In Grothendieck's way of doing things, in particular, we can observe, firstly, the introduction of a web of incessant *transfers*, *transcriptions*, *translations* of concepts and objects between apparently distant regions of mathematics, and, secondly, an equally incessant search for *invariants*, *protoconcepts and proto-objects* behind that web of movements. In their technical definitions, sheaves and schemes incarnate both flux and repose. Beyond sheaves as singular objects, the 'protogeometry' that underlies certain *classes* of sheaves then gives rise to the *Grothendieck topoi*.

Grothendieck topoi (1962) are categories of sheaves issuing from certain 'natural' abstract topologies.[95] Topoi, which are something like *parallel universes* for the development of mathematics, are categorical environments sufficiently vast for the development of an entire sophisticated technology of the relative to be possible. Generalizing the action of certain groupoids on the fibers of a sheaf, Grothendieck seeks *to move the topoi* (environments that are no longer just set-theoretical but topological, algebraic, differential, combinatorial, etc.) and study, in a generic fashion, the actions of various functors on enormous

---

95 In categories with certain good properties of compositionality and covering, an abstract topology (*Grothendieck topology*) can be defined by means of (sub)collections of morphisms that are 'well matched' with one another. The categories of *presheaves* (categories of functors to values in the category of sets) verify those good properties of compositionality and covering, and abstract topologies can be defined there. Grothendieck topoi issue from categories of presheaves that are 'situated' around a given abstract topology. (Those categorical environments are also called *sites*.) A simplification of Grothendieck topoi is provided by Lawvere's elementary topoi (1970), where the abstract topologies (by means of Yoneda's lemma) can be easily described thanks to *a single* endomorphism of the subobject classifier, which registers the algebraic properties of a closure operator.

classes of topoi. The results come without delay, and *it is in the generic geometrical realm of topoi that certain cohomological obstructions disappear*: it is there that Grothendieck and his school could develop the étale cohomology that would allow Deligne to resolve the Weil conjectures. In topoi, objects are no longer 'fixed' but 'unfold through time'. We are dealing here with *variable sets*, whose progressive parametric adjustments allow us to resolve a multitude of obstructions that seemed irresolvable from within a 'punctual', classical or static mathematics. From this alone, one can intuit the enormous philosophical impact that such a *relative mathematics* might have – a mathematics attentive to the phenomenon of *shifting*, but with the capacity to detect invariants behind the flux, a mathematics that goes against the grain of supposedly ultimate foundations, absolute truths, unshakable stabilities, but which is nevertheless capable of stabilizing *asymptotic webs* of truth.

In Grothendieck's work, objects tend to be situated over certain 'bases' (the sheaf over its underlying topological space, the scheme over its spectrum), and many important problems arise when *base changes* are carried out. 'Relative' mathematics then acquires a great mathematical incisiveness, in inquiring as to which properties are transferred in the effectuation of base changes (*descent theory*: the search for the conditions under which one is able to carry out transfers; and its counterpart, the detection of the conditions of obstruction in base changes). Conditions of coherence and abstract gluing arise quite naturally from these processes of transference/translation – conditions that, it seems, can be easily defined only in the context of Grothendieck topoi. In particular, Zariski's 'site', which allows us to articulate the Weil conjectures, is replaced by Grothendieck's 'étale site',[96] in which is constructed – following a general procedure that we will come back to in the *Tohoku*, according to which certain categories of sheaves give rise to natural cohomology groups – the étale cohomology that Deligne will later need in order to resolve the conjectures.[97]

In reality, the conceptual dynamics of topoi far surpass the theory's first technical objectives, as brilliant as they were. Indeed, surfacing behind the Grothendieck topoi Lawvere's elementary topoi, where we see that the number-theoretic, algebraic, topological and geometrical considerations advanced by Grothendieck also possess surprising *logical* counterparts.[98] As we will see in chapters 5 and 7, as we approach Lawvere and Freyd, the categories

---

96 *Étale*: smooth, without protuberances (the term comes from a poem of Victor Hugo's, about an '*étale*' sea). Grothendieck's metaphorical use of '*étale*' condenses the idea of the *nonramified*, where Grothendieck combines, once again, some of the central ideas of Galois and Riemann's: extensions of nonramified fields (Galois's separability) and nonramified Riemann surfaces, enveloped in a generic unifying concept.

97 See P. Deligne, 'Quelques idées maîtresses de l'oeuvre de A. Grothendieck', in *Matériaux pour l'histoire des mathématiques au XXe siècle*, Séminaires et Congrès 3, Société Mathématique de France, 1998, 11–19.

98 Oddly, Grothendieck, who explored almost every field of mathematics with tremendous penetration, hardly bothered with mathematical logic. That disquieting logico-mathematical separation – by one of the two or three major mathematicians of the twentieth century – should give logic-centered philosophers of mathematics much to think about.

and allegories that situate themselves between cartesian categories and topoi encode an entire legion of intermediary logics, whose *relative web* reflects a good part of the greater mathematical movements that are based on them. *Base changes in underlying logics* thus give rise to a complex landscape – we could call it *relative logic* – that allows us to return to the historical origins of mathematical logic (Peirce's 'logic of relatives') and reinterpret, in a new light, many of the problematics concerning foundations, which were taken up in a conventional way by analytic philosophy.

The Grothendieckian attention to the movement of mathematical concepts and objects is accompanied by an oscillating search for *archetypes* for mathematical reason and imagination. Between the *One* (the 'form') and the *Many* (the structures: schemes, topoi, etc.), Grothendieck discovers and invents[99] suitable invariants of form: the cohomologies. Although the homology and cohomology groups for algebraic topology tend to satisfy certain conditions of univocity, the possibilities for cohomological invariants multiply as we move through algebraic geometry (Hodge, De Rham, crystalline, étale, l-adic, etc.). It is for this reason that Grothendieck proposes his *motifs*,

---

99 We will deepen, in section 4.2, the dialectic of discovery and invention in Grothendieck, a dialectic that *cannot be reduced* to either of its two poles; and we will study with greater care, in part 3, the fact that a position as much realist ('discovery') as idealist ('invention'), is indispensable in advanced mathematics, once the latter goes beyond a certain *threshold of complexity* for structures, languages and the transits/obstructions at stake.

deep generic structures underlying distinct cohomologies. Reading Grothendieck's own words is well worth the trouble, since we will be taking up several ideas from the following quotation throughout this essay:

> This theme [that of motifs] is like the *heart* or the soul, the most hidden part, the most concealed from view, of the schematic theme, which is itself at the heart of a new vision. [...] Contrary to what happens in ordinary topology, [in algebraic geometry] we find ourselves faced with a disconcerting abundance of different cohomological theories. One gets the very clear impression that, in a sense that at the beginning remains somewhat vague, all of these theories should 'turn out to be the same', that they 'give the same results'. It is to be able to express this intuition of 'kinship' between different cohomological theories that I have extracted the notion of a '*motif*' associated with an algebraic variety. By this term I mean to suggest that what is at stake is the 'common motif' (or the 'common *reason*') underlying this multitude of different cohomological invariants associated with the variety, thanks to the entire multitude of cohomological theories that are possible a priori. These different cohomological theories would be something like different thematic developments, each in its own 'tempo', 'key' and 'mode' ('major'

or 'minor'), of the same 'basic motif' (called '*motivic* cohomological theory'), which would be at the same time the most fundamental, or the most 'fine', of all these different thematic 'incarnations' (that is to say, of all these possible cohomological theories). And so, the motif associated with an algebraic variety would constitute the 'ultimate' cohomological invariant, the cohomological invariant '*par excellence*', from which all the others [...] would be deduced, as so many different musical 'incarnations' or 'realizations'. All the essential properties of '*the* cohomology' of the variety would already be 'read off' (or 'heard in') the corresponding motif, so that the properties and structures familiar to the particularized cohomological invariants (*l*-adic or crystalline, for example), would simply be the faithful reflection of the properties and structures *internal to the motif*.[100]

Both homologies (mathematical constructions that help us resolve 'the discrete/continuous aporia' [Thom] and that consist of chains of abelian groups with which one captures ample information about the topological object

---

100 Grothendieck, 'Prélude', in *Récoltes et Semailles* 45–6 (quotation marks and italics are the author's). The (conceptual, mathematical, stylistic, methodological, phenomenological) richness of this paragraph will give rise to many reflections in our work. For the moment, it is enough to underscore the movement between the One and the Many, the tension between the 'ultimate' and the differences, the problematic of fidelity and variation, the dialectic between the internal and the external, the modal spectrum of possibilities and realizations, the interlacing of vagueness and precision, the grafting of *corazón* and *razón* (heart and reason), the aesthetic equilibrium.

under investigation) and cohomologies (dual constructions involving more familiar set-theoretical limits [products, pullbacks, etc.]) become, thanks to Grothendieck, some of the 'most powerful [mathematical] instruments of the century'.[101] At the end of his research at the IHES, after his work on schemes and topoi, Grothendieck envisioned the difficult and ambitious *motivic program*. Retiring from the world of mathematics and ceasing to publish, the major lines of development of Grothendieck's program went on to circulate as manuscripts, and many of his suggestions were considered excessively vague.[102] Nevertheless, Voevodsky introduced *motivic cohomology* (1990–2000), a contribution that, in part, answered Grothendieck's expectations, and that won him the Fields Medal (2002). Instead of working, as in algebraic topology, with algebraic surgeries on space (singular cohomology, ring of cohomology groups), Voevodsky proposed a more delicate collection of *surgeries on an algebraic variety*, introducing new forms of topology for algebraic objects (fine Grothendieck topologies over sites of schemes),

---

101 Grothendieck, *Récoltes et Semailles*, 43.

102 The same could be said of another very influential 'vision' of Grothendieck's, the 'moderate program' that he sketched out in his 1983 *Esquisse d'une programme*. Grothendieck sought new forms of topology, which would turn out to be natural and would *smooth over* the singularities that a set-theoretical topology must endure (replete with artificial examples coming from analysis). Grothendieck had the intuition that a sort of deconstruction ('*dévissage*', *Esquisse*, 25) of stratified collections of structures would be tied to the discovery of a 'moderate topology'. Amid the developments of the 'moderate program' one can find *tame model theory* or 0-minimality in contemporary model theory – another unsuspected resonance of Grothendieck's ideas with logic.

and defining a sophisticated *concrete category* for the $H(V)$ homologies functorially associated with varieties V. A *central trunk of cohomologies* had then begun to 'surface', concordant with Grothendieck's extraordinary mathematical intuition.

4.2 METAPHORS, METHODS, STYLE

In this section we analyze some of the major metaphors that Grothendieck himself used in explicating his modes of creation and work methods, and we observe some of the *resonances* we encounter between Grothendieck's mathematical *production* (in its imaginative phase no less than its definitional and theorematic phase), the mathematician's *reflection* on that production, and the formal *expression* of that reflection. All of these resonances constitute what could be called the peculiar *style* of Grothendieck.

The metaphors of the 'hammer' and of the 'rising tide' preside over much of Grothendieck's conceptual vision.[103] For Grothendieck, a *problem* can be imagined as a sort of 'nut', whose hard shell has to be penetrated in order to get to its 'soft flesh'. In Grothendieck's conception, there are two essentially distinct strategies for opening the shell: hitting it with a hammer and chisel – sometimes slipping and sometimes smashing the inside to pieces

---

103 Grothendieck, *Récoltes et Semailles*, 552-3.

along with the shell – and immersing it in a liquid ('the tide') in such a way that, after weeks or months, its exterior softens and opens up 'with a squeeze of the hands [...] like a ripe avocado'. The first strategy (yang) aims to *resolve* the problem; the second strategy (yin) aims to *dissolve* it. Through an adequate *immersion in a natural, ambient medium*, the solution should *emerge* within a *generic* landscape that outstrips the particular irregularities of the shell. The metaphors capture a precise mathematical methodology that Grothendieck had constantly put into practice over at least thirty years: immersing a problem in an appropriate general category (K), performing a profound labor of conceptual and definitional *prescission*[104] inside that category, decomposing examples and objects inside that general frame, and proceeding finally to the study of the correlations, transits and osmoses within the category. After an incessant abstract (de)construction ('*dévissage*'), the problem can be resolved with the greatest possible *softness* ('a ripe avocado'), without blows and without artificial ruses, as the direct testimony of Deligne indicates.[105]

Going further still, Grothendieck's strategy of the 'rising tide' goes on to place questions, notions and points

---

104 *Prescission*, in Peirce's sense, at once *cuts and specifies* the boundaries of the entity under analysis.

105 Deligne, '*Quelques idées maîtresses...*', 12.

of view at the center of mathematical attention, above and beyond the resolutions themselves:

> More than anything, it's really through the discovery of new *questions*, and likewise new *notions*, and even new *points of view* – new 'worlds', in fact – that my mathematical work has turned out to be fruitful, rather than through the 'solutions' that I have contributed to questions already posed. This very strong drive, which has carried me toward the discovery of good questions, rather than toward answers, and toward the discovery of good notions and statements, much more than towards proofs, is another strong 'yin' trait in my approach to mathematics.[106]

Behind a problem, Grothendieck always seeks out the wellsprings (the sources) of natural questions associated with the problem. What is at issue, therefore, is a vision

---

106 Grothendieck, *Récoltes et Semailles*, 554. Of course, such a paragraph can only be appreciated from a great height, seeking to clarify the most salient movements of the topography. We must not forget the (literally) thousands of pages that Grothendieck devoted to 'answers' and 'demonstrations' in his major fields of production: 'tensorial products and nuclear spaces, cohomologies of sheaves as derived functors, K-theory and the Grothendieck-Riemann-Roch Theorem, emphasis on work relative to a base, definition and construction of geometrical objects via the functors to that which must represent them, fibered categories and descent, *stacks*, Grothendieck topologies (sites) and topoi, derived categories, formalisms of local and global duality (the "six operations"), étale cohomology and cohomological interpretation of L-functions, crystalline cohomology, "standard conjectures", motifs and the "yoga of weights", tensorial categories and motivic Galois groups' (following a 'brief' list of contributions, in P. Cartier et al., *The Grothendieck Festschrift* [Basel: Birkhäuser, 1990], vol. 1, viii). As Dieudonné points out (ibid., 14), 'there are few examples in mathematics of so monumental and fruitful a theory, built up in such a short time, and essentially due to the work of a single man'.

of the *foundations* of mathematics that differs radically from the one proposed by set theory. Grothendieck's 'reading' is a transversal one, in which an ultimate base is of no importance. What is under investigation, instead, is the base's movement (its *shifting*),[107] and what matters, more than an accumulative resolution of knowledge, is the mobile interlacing of natural questions underlying the solutions.[108]

In fact, it is not even a question of a 'reading' in Grothendieck, but rather a *listening*. An articulation between *images, intuition and ear*, as opposed to other merely formal manipulations of language, seems to be fundamental for him. In addition to the metaphor of the nut and the rising tide, another of Grothendieck's central metaphors is, in fact, the image of the creative mathematician attending to 'the voice of things'.[109]

---

107 For Merleau-Ponty, the 'height of reason' consists in *feeling the ground slip away*, detecting the movement of our beliefs and our supposed knowledge: 'every creation changes, alters, elucidates, deepens, confirms, exalts, recreates or precreates all the rest' (M. Merleau-Ponty, *Notes des cours de Collège de France [1958–59, 1960–61]* [Paris: Gallimard, 1996], 92). In *L'oeil et l'esprit* (Paris: Gallimard, 1964), Merleau-Ponty describes the body as operating in the domains of knowledge as a 'sheaf of functions interlacing vision and movement'. Through incessant levels of self-reference, the sheaf permits the conjugation of inner and outer, essence and existence, reality and imagination. And, moreover, it is in the murky and antinomic frontiers of such apparent contradiction that the sheaf gives rise to invention and creation. We shall return, in part 3 (chapter 10) to certain connections between Grothendieck, Merleau-Ponty and Rota, as regards the apparently fundamental opposition between invention and discovery in mathematical philosophy.

108 Recall, here, the similar position of Lautman, who pointed to an 'urgency of problems, behind the discovery of their solutions' (pp66–8, above).

109 Grothendieck, *Récoltes et Semailles*, 27.

The 'hidden beauty of things'[110] appears to be the hidden beauty of mathematical structures, an intrinsic beauty that the mathematician *discovers by means of the extrinsic invention* of sufficiently expressive languages. And so, in Grothendieck's perspective, mathematical structures appear in the phenomenological spectrum of the world, and so they are discovered – but these are discoveries that can only be made by inventing, in an almost synchronic dialectic, adequate *representations* of the structures in question. The (musical, cohomological) metaphor of the *motif* itself shores up the idea that there exist *hidden germs of structuration*, which a good 'ear' should be able to detect. And so Grothendieck's motifs appear to be *already* present in the dynamic structure of forms, independent of their future discoverers (Voevosdky, Levine, Morel, etc.), whose work would consist essentially in creating the adequate languages, the theoretico-practical frameworks, and the sound boxes required to register their vibrations. Again, it is instructive to listen to Grothendieck himself:

> The structure of a thing is not in any way something that we can 'invent'. We can only patiently, humbly put it in play – making it known, '*discovering it*'. If there is inventiveness in this work, and if we happen to perform something like the work of the blacksmith

or the tireless construction worker, this is not at all to 'fashion' or 'build' structures. They do not wait for us in order to be, and to be exactly what they are! It is rather to express, as faithfully as we can, these things that we are in the midst of discovering and sounding out – that reticent structure toward which we try to grope our way with a perhaps still-babbling language. And so we are lead to constantly '*invent*' *the language* that can express, ever more finely, the intimate structure of the mathematical thing, and to 'construct', with the help of that language, thoroughly and step by step, the 'theories' charged with accounting for what has been apprehended and seen. There is a continual and uninterrupted back-and-forth movement here, between the *apprehension* of things and the *expression* of what has been apprehended, by way of a language that has been refined and recreated as the work unwinds, under the constant pressure of immediate needs.[111]

The incessant recreation and invention through the discipline's unwinding environs, the step-by-step construction, and the groping expression all evince Grothendieck's eminently dynamical and dialectical perception. What we are dealing with here, in effect, is a 'continual weaving'

---

111 Ibid., 27. The quotation marks and italics are Grothendieck's.

in mathematical thought, back and forth with a probing instrumentarium, whose ontological and epistemological categories *cannot be stabilized in advance*, independently of the (practical, historical) action of the discipline. In that ubiquitous transit in which mathematics returns to itself, in that always-moving and often enigmatic sea, Grothendieck's *manner* nevertheless provides a profound *orientation* and a surprising relative *anchorage*.

In effect, Grothendieck relies on multiple mathematical *methods* in order to maintain an orientation within the variable scenery that looms up ahead. Above all, a persistent *ascent and descent* allows him to overcome the obstructions that lay in wait in local and excessively particularized labyrinths. The ascent to the general is never, in Grothendieck, a gratuitous operation, but is controlled by certain crucial modes of mathematical practice: the insertion of a particular local situation (object, property, example) into a universal, global environment (category) – with subsequent osmoses between manifestations of the singular and forms of the continuous; the plural construction of webs and hierarchies so as to collate the particular within wider relational universes; the discovery of proximities in a topography with clearly defined elevations and projections of various types. Generalization is then a weapon of contrast, a method for elevating vision, that helps us to orient ourselves in a complex terrain.

On the other hand, Grothendieck's 'manner' is entirely governed by a ubiquitous (conceptual, linguistic, technical) *dialectic*. From the most vague (the yin and the yang) to the most precise (functorial adjunctions), passing through incessant tensions between polar regions of mathematics, Grothendieck's thought comes and goes without the slightest rest. Many of his great technical constructions – nuclear spaces, K-theory and generalized Riemann-Roch, cohomologies, schemes, topoi, motifs – straddle apparently distant mathematical nuclei, culminating in 1983 in the *dessins d'enfants* (children's drawings), which propose strange combinatorial invariants for number theory, by way of surprising mediations on analysis (Riemann surfaces) and algebra (Galois groups). The dialectic functions on multiple levels, from the vague and imaginary to the technically restrained, in a 'vast counterpoint – in a harmony that conjugates them'.[112]

Restricting the dialectic to the subdefinition of *transits and obstructions* within mathematical activity, the genericity of mathematical concepts and objects (à la Grothendieck) gives rise to other original concretizations in the spectrum of the *arbitrary* – arbitrariness being understood as a simultaneous *topos* of mediation ('arbiter', transit, continuity) and opposition ('arbitrary', imposition, discretion).[113]

---

112 Ibid., 23.

113 I owe this beautiful dialectical and etymological reading of the 'arbitrary' to Roberto Perry and Lorena Ham, in whose work it gives rise to a complex 'therapeutic of the arbitrary'.

In effect, a generic entity *combines* its implicit definability within a horizon of possibility with its explicit concretization within a horizon of stratification (Desanti),[114] so as to project its abstract capacity for transit (amid *possibilia*) onto the concrete panorama of impositions that it encounters (in the hierarchies of the actual). Whence a third method emerges, one that is very much present in all of Grothendieck's works, in the 'art' of making mathematics. What we are dealing with here is, indeed, a new *ars combinatoria*, which proposes to explicate, in four well-defined steps, mathematics' *unity in multiplicity*, which is to say, 'the very life and breath'[115] of the discipline: the incessant stratification of mathematical activity; the ramification of ambient categories of interpretation; the recursive deconstruction ('*dévissage*') of the concepts at stake, throughout the many available categorical hierarchies; the framing of relational interlacings (diagrams of transferences and obstructions) between realized deconstructions.

Grothendieck's *manner*[116] – a mixture of vertical

---

114 J.-T. Desanti, *Les idéalités mathématiques* (Paris: Seuil: 1968).

115 Grothendieck, *Récoltes et Semailles*, 16.

116 In the artistic theory of the sixteenth and seventeenth centuries, *maniera* appears at the nucleus of critical discussions on great painter's 'ways of doing things' (of inventing, creating). With the degeneration of *maniera* into *mannerism*, the notion of *style* later emerged as a conceptual substitute for capturing the major categories of the history of art (baroque, classical, romantic, etc.). In part 3 of this work, we will take up the problematic of how we may try to define – intrinsically, and not just diachronically, as we have done so far – some of the great demarcations of mathematical styles: classical, modern, romantic, contemporary. The *maniera* of Grothendieck opens important channels for attempting to approach such intrinsic demarcations. It is something that we have already begun in our chapter 1 – with conditions 1–5 in terms of which Lautman investigated modern mathematics, and with conditions 6–10, closely tied

(ascent/descent), horizontal (dialectics/polarities) and diagonal (reflections between stratified hierarchies) weavings – gives rise to a *style* that allows us to naturally express these ways of making mathematics. By 'style', here, we will understand the superposition of webs of 'inscriptions' (recalling the *stylus* with which the Babylonian tablets were inscribed) and the intermeshing 'gears' between the webs, in three fundamental mathematical registers: (*i*) the initially vague invention, (*ii*) the subsequent delimitation of that vagueness and consequent demonstrative expressiveness, (*iii*) the critical reflection on the demonstrative body whose elaboration has been made possible.[117] From this perspective, Grothendieck's style is of the greatest interest because it amply extends through the three registers,[118] and, in the epistemological sense of the word 'style', it effects a genuine 'saying-thinking' conjugation (*lexis*)[119] in each field. Grothendieck, in fact, describes his 'particular genius'[120] as a capacity for introducing great new themes and unifying *points of view* amid diversity; and

---

to Grothendieck's *maniera*, regarding which we have approached the 'contemporary' – which we shall study more carefully, however, in Chapter 11.

117 These three registers correspond to forms of the three Peircean categories: (*i*) firstness and abduction; (*ii*) secondness and contradistinction; (*iii*) thirdness and mediation. The *logic of scientific inquiry*, extensively studied by Peirce, links together precise modes of transforming information between these three categories. We will return to these questions in chapter 10.

118 (*i*) : Correspondence; (*ii*): EGA, SGA, articles; (*iii*) *Récoltes et Semailles*. In each register, the studious reader can count on hundreds of pages for developing her observations.

119 See the entry on 'Style' in B. Cassin, ed., *Vocabulaire européen des philosophies* (Paris: Seuil, 2004), 1,226.

120 *Récoltes et Semailles*, 15.

his skill in 'saying' – that is, the *rich traces of his style* – is a vital instrument for his unitary capacity for 'thought'.

The correspondence with Serre reveals Grothendieck's indomitable energy, his potent mathematical inventiveness, his staggering capacity for abstraction and concentration, but also his doubts and errors, his desire to 'cultivate himself' with the help of his correspondent's enormous breadth of mathematical knowledge, as well as the melancholy twilight of his great critical brilliance (in the final letters of 1985–1987, around the time of *Récoltes et Semailles*).[121] The enormous technical complexity of the correspondence[122] does not prevent us from being able to detect many moments in which Grothendieck's ideas continually *emerge over the course of days*. From the point of view of style, let us underscore, among other things, the 'cohomology deluge' (1956) that gave rise to the article 'submitted to Tannaka for the *Tohoku*', the writing of which is contrasted with 'Weil's truly intimidating demonstrations';[123] Grothendieck's incessant preoccupation with defining 'clearly natural' concepts and ideas, which distinguish his *manner* of making mathematics from other artificial practices;[124] the presence of a 'plausible

---

121 P. Colmez, J.-P. Serre, eds., *Correspondance Grothendieck-Serre* (Paris: Société Mathématique de France, 2001).

122 Extramathematical remarks make no appearance in the correspondence, with the exception of a brief Grothendieck-Serre-Cartan exchange (1961) about the problems that military service is causing for young mathematicians. Ibid., 121–8.

123 Ibid., 38, 49.

124 Ibid., 111.

yogical reason' that helps to specify and orient certain general speculations;[125] the explosion of gigantic 'river letters [*lettres fleuve*]' in 1964, at the moment of the invention of the Grothendieckian ideas about *l*-adic cohomology (which ten years later would lead to the resolution of the Weil conjectures). The nourishing correspondence – which would be accompanied by hours (!) of telephone conversations between Grothendieck and Serre, an entirely original way (and one incomprehensible to mere mortals) of sharing high mathematics – allows us to glimpse a *gestating* mathematics: mobile, imbued with a remarkable vigor, replete with approximations between distant concepts, in a process of refinement, with a perpetual correction of (natural) errors between one letter and the next. Being able to spy on the blacksmith at his forge, we can now *see* a *real* mathematics, so far from that which the predominant currents of analytic philosophy have taught us to appreciate.

As a counterpart to his rare talent for 'discovery' and his facility for training the ear to listen to the 'voice of things', Grothendieck also had a rare sensitivity for the use of language, which he succeeded in concretizing in *terminological* 'inventions' as explosive as his mathematical inventiveness itself. An integral part of Grothendieck's

---

125 Ibid. The comical neologism 'yogical [*yogique*]' invokes Grothendieckian *yoga* (the 'vague' interlacings of yin-yang dialectics) and is contrasted with a supposedly formal and precise *logic*.

mathematical style, the terminology, in fact, formed an entire universe in itself. With the same fascination that Proust expresses in 'Place Names: The Name',[126] Grothendieck declares that 'one of my passions has constantly been *to name* the things that discover themselves to me, as a first means of apprehending them',[127] and points out that

> from a quantitative point of view, my work during those years of intense productivity has principally concretized itself in some twelve thousand pages of publications, in the form of articles, monographs and seminars, and by means of hundreds, if not thousands, of new notions that have entered into the common patrimony with the same names that I gave them when I delivered them. In the history of mathematics, I believe I am the one who has introduced the greatest number of new notions into our science, and the one who has been led, by that very fact, to invent the greatest number of new names in order to express those notions with delicacy, and in as suggestive a fashion as I could.[128]

---

126 'Du côté de chez Swann [Swann's Way]', in M. Proust, *À la recherche du temps perdu* (Paris: Gallimard, 1997). The latest edition, by J.-Y. Tadié, collates all the variants of the original manuscript and thereby thoroughly explores the place of Proustian inventiveness. [Tr. C. K. Scott Moncrieff & T. Kilmartin, *In Search of Lost Time* (New York: Modern Library, 1998).]

127 *Récoltes et Semailles*, 24.

128 Ibid., 19.

The 'delicacy' of the name, the 'suggestiveness' of metaphors, the profusion of names that go on to modify a 'common patrimony' are not contemplated in the treatises of mathematical philosophy. Even in schools ensnared by a fascination with language, no study of the language of *real mathematics* has been undertaken, even though lengthy disquisitions have been elaborated on a language supposedly capable of supplanting mathematics itself. It is not our intention to go further into mathematical language, but it must be pointed out that Grothendieck's terminological inventiveness ought to be explored, elsewhere, with the care that it deserves.

4.3 THREE PARADIGMATIC TEXTS: *TOHOKU*, EGA, CARTAN SEMINAR

In this final section, we will review three specific texts by Grothendieck, aiming to unveil some of the great lines of tension in Grothendieck's production, as underscored in the first section, as well as some of the methodological, metaphorical and stylistic procedures indicated in the second. We will look at three paradigmatic texts:

1. the *Tohoku* (published in 1957, but which *emerges* from 1955 on, as we see mentioned in the correspondence with Serre);

2. the first chapters of the great treatise, *Elements of Algebraic Geometry* (published in 1960, but already in gestation

a few years earlier, as we can also deduce from the correspondence with Serre);

3. the extraordinary series of expositions in the Cartan Seminar of 1960–61, concerning themes in 'analytic geometry' (understood, in Cartan and Dieudonné's manner, as the geometry of analytic functions).

As Grothendieck points out at the beginning of the article, the *Tohoku* aims to explicate a 'common framework' that would allow us to nourish the 'formal analogy' between the cohomology of a space with coefficients in a sheaf and the series of derived functors of a functor on a category of modules.[129] The work was written at the very moment when the notion of sheaf came to constitute an indispensable component of mathematical investigation,[130] an investigation during the development of which Grothendieck made a point of mentioning his 'conversations' with Cartan, Godement and Serre.[131] Now, from the very beginning of the *Tohoku*, vague 'analogies' and dynamic forms of mathematical creativity appear in

---

129 Grothendieck, 'Sur quelques points d'algèbre homologique', 119.

130 The concept of mathematical sheaf emerges in the work of Jean Leray, in a course on algebraic topology in the Oflag XVII-A* (1943–5), a series of notes in the *Comptes Rendus de l'Académie des Sciences* (1946), courses on spectral successions at the *Collège de France* (1947–50), and reaches its definitive development in the Henri Cartan Seminar at *l'École Normale Supérieure* (1948–51). (Oflag XVII-A was an unusual German prisoner-of-war camp in which officers were imprisoned during WWII; Jean Leray directed the camp's university.) A sheaf is a type of mathematical object that allows for the global gluing of whatever proves to be coherently transferable in the local. Certain mathematical objects can then be better understood, thanks to a logic of *neighborhoods and mediations* over a continuous space (getting away from yes-no binarisms), and to the natural actions of groupoids in the fibers of the sheaf.

131 Ibid., 120–1. Grothendieck also spoke of the 'interest' of his colleagues.

the text, and Grothendieck immediately proceeds to *invent* the language (additive and abelian categories) and *discover* the richness of the mathematical structures (sheaves, injective objects via products and infinite sums, actions of a group) that explicate the initial 'formal analogy'. In particular, the actions of the groups at stake *are stratified* in a precise hierarchy of levels – action on a space X (first), action on a sheaf O of rings over X (second), action on a sheaf of modules over O (third) – thereby concretizing one of the typical forms of the Grothendieckian process. This strategy gives rise to the article's various parts: (I) abelian categories (*language*); (II) homological algebra in abelian categories and (III) cohomology with coefficients in a sheaf (*structures*); (IV) *Ext* calculi for sheaves of modules and (V) cohomologies with spaces of operators (*transferences and actions*).

The 'grand vision' of category theory proposed by Grothendieck in section (I) remains extraordinarily fresh after fifty years.[132] It is dealt with in the lengthy introductory section of the text, where Grothendieck establishes the three clear levels of categorical thought – morphisms, functors, natural transformations (called 'functorial

---

[132] We could consider the present essay (originally published in 2007), to a large extent, as an homage to the *Tohoku*, fifty years after its publication. It would perhaps be too much of an exaggeration to describe contemporary mathematics as a series of footnotes to the work of Grothendieck, but the exaggeration would have an unquestionable grain of truth to it.

morphisms' in the text).[133] In the same section, he also introduces his additive and abelian categories, compares existence axioms for infinite products, establishes the existence of sufficient injectives (via diagrams, generators and products) and develops the quotient categories.[134] In the process, Grothendieck produces some remarkable *examples*, placing them like little crystalline gems in the weave of ascent and descent between the universal and the concrete. As instances of additive non-abelian categories,[135] for example, we are shown separated topological modules, filtered abelian groups, and holomorphic fiber spaces over a 1-dimensional manifold. In this way, the *topographical richness* of Grothendieck's mathematical thought is never gratuitous, its ascent is never driven by an artificial impulse to generalize, and it always contemplates a vivid landscape of *specific* hills and valleys. This is something that receives further corroboration with the comparison of various axioms for infinite products in terms of the classes of concrete

---

133 Ibid., 124.

134 Grothendieck gives the name 'Serre's module language C' to the idea of variation over the base (ibid., 137). Given that what is at issue here is one of the central ideas governing a good deal of 'relative mathematics' and the Grothendieckian 'tide', it is very interesting to observe how, *in the very movement of the idea's emergence*, Grothendieck sees Serre as the 'creator' of 'modulation through C'. This is one more proof of the incessant *contamination* of mathematics, with every sort of *residue* in a web of mixtures and impurities, that falls outside of analytic philosophy's modes of observation.

135 Ibid., 127.

categories that they distinguish,[136] with a 'fun example'[137] having to do with cohomologies over an irreducible space, with a profound example dealing with sheaves of germs and differential forms over holomorphic manifolds,[138] and with further examples of functorial manipulations that allow us to reconstruct earlier arguments in which Grothendieck utilizes the machinery of Cech coverings.[139]

The 'common framework' emerging from the *Tohoku* – constructed in order to allow for the study of *natural interlacings* between algebraic geometry, topology, complex analysis and cohomological calculi – went on to modify the mathematical landscape. Focusing his efforts on a *pivotal* mathematical concept/object (the mathematical sheaf), defining the general environments in which sheaves can be studied in their unity/multiplicity (abelian categories), and putting this entire instrumentarium at the service of comprehending the deep forms of structures (the cohomologies), Grothendieck brings about not only a

---

136 Ibid., 129. The category of abelian groups, its dual category of compact topological abelian groups, and the category of sheaves of abelian groups over a given topological space all make appearances. The instruments of transfer (the Pontrjagin duality) and of *integral* leaps from one level to another (sheaves) are put in the service of a *differential* understanding between categories – germs of a very abstract, contemporary, differential and integral calculus.

137 'Un exemple amusant', ibid., 160. 'Fun' is not well represented in 'formal' mathematics, but is, without a doubt, among the important motors of the creative mathematician, something which, of course, appears to be indiscernible in both 'normal' mathematics (the series of texts published in the community) and in philosophical discussions of this normal tradition.

138 Ibid., 165–6.

139 Ibid., 161, 213. The interlacing between functorial descriptions and Cech-style coverings can probably be seen as the very origin of the Grothendieck topologies.

'Copernican turn' in mathematics, but a genuine 'Einsteinian turn', if we allow ourselves to stretch the metaphor a little.[140] Grothendieck's vision even comes to *transcend* the framework that he himself elaborates, when, in a brilliant premonition, he observes that 'it would be a good idea to provide a treatment of "noncommutative homological algebra"' in a context of functors and categories that envelopes the theory of fibrations and the extensions of Lie groups[141] – a startling anticipation of fragments of Connes's program of noncommutative geometry, which we will review in chapter 6.

Like the *Tohoku*, the *Éléments de Géométrie Algébrique* (EGA) incorporate the most visible characteristics of Grothendieck's procedure. Although the metaphorical,

---

140 For his part, Peirce brings about what could be called an 'Einsteinian turn' in philosophy. Of course, although Peirce preceded Einstein and the label is therefore paradoxical, the universal Peircean semiosis and its associated construction of relational invariants is precisely fitted to the 'revolution' in modern physics that Einstein would bring about just one decade later. In Peircean semiosis, subject and object are considered not as monadic predicates but as relational *webs* of various signs, inserted in scaffoldings of reference subject to a perpetual dynamism ('unlimited semiosis'). In that dynamism of *relative movements*, even the observation of an object can undergo modification. Peirce therefore tries to find invariants in that complex relational flux: the 'Einsteinian turn' of his philosophy seeks (and finds) what we could call the philosophical invariants of a general logic of relations and higher-order logic. The relativity of perspective, the unlimited dynamism of interpretation and the modification of interpretants are some of the great conquests of the Peircean system – conquests that the twentieth century repeatedly corroborates in the most diverse guises. Nevertheless, with his system's permanent processes of *reintegration* and *gluing*, Peirce overcomes the extreme relativism into which certain rehabilitations of the ephemeral and the local in the last stages of the twentieth century can be seen to have led. We will take these ideas regarding a supposed 'Einsteinian turn' in philosophy (Peirce) and an 'Einsteinian turn' in mathematics (Grothendieck) further in part 3. If those approximations are more or less correct, the philosophy of mathematics should, in turn, undergo a *considerable turn*.

141 Ibid., 213.

analogical and stylistic levels are reduced to a minimum in the EGA (an asepsis owing, to a large extent, to Dieudonné's steely coauthorship), they incorporate a 'grand vision' (a mathematics *in action* that would provide the bases from which to attack the Weil conjectures, 'a labor scarcely undertaken'[142] in 1960), an *open* way of propelling and presenting the discipline,[143] a clear global landscape with some well-delineated local techniques,[144] a general way of doing mathematics within contexts (*categories*) sensitive to the transit/obstruction of information (*functors*, natural transformations),[145] and a persistent search for the *natural* notions governing those osmoses.[146]

---

142 EGA, I, 9.

143 'All of the chapters are considered to be open', ibid., 6. Observe, indeed, how chapter 0 ends with the phrase '*A suivre* [to be continued]' (ibid., 78), something rather uncommon in the mathematical literature, where texts are usually presented as 'finished'. Mathematics *in gestation* always ends by emerging (more obscurely than it sometimes seems) in Grothendieck's works.

144 Given two algebraic varieties X, Y (or, more generally, given two schemes), the study of the properties of a problem P in a neighborhood of $y \in Y$ is approached by way of its transformations/obstructions through a (proper) morphism between X and Y, following precise steps in the analysis of the problem: introducing the study of an adequate local ring A over y; reducing that study to the artinian case (with which one moves to a 'greater understanding of the problem, which on this level is of an 'infinitesimal' nature' [ibid., 8]); effecting adequate passages by means of the general theory of schemes; permitting the discovery of algebraic extensions of A (the primordial task of algebraic geometry) by means of adequate multiform sections of the schemes.

145 Several sections of chapter 1 ('The Language of Schemes') answer precisely to the study of schemes from the *categorical* point of view of the preschemes that envelop them: products (§3); subobjects (§4); separability conditions (§5); and finitude conditions (§6). As in the *Tôhoku*, sophisticated examples are introduced (in polynomial rings, ibid., 139) in order to distinguish, by means of suitable models, the various conditions of separability.

146 'The usual constructions suggested by geometrical intuition can be transcribed, *essentially in a single reasonable manner*, into this language [of schemes]' (ibid., 9, our italics). Grothendieck once again expresses one of his deep convictions: behind the plurality of structures and signs, the construction (invention) of an adequate language should allow for the *naturalness* of certain 'One'-structures (discovery) from which

As we saw in section 4.1, schemes let us elaborate an important unification of the visions of Galois and Riemann, a global fact that gets countersigned locally when Grothendieck and Dieudonné explain that the 'lag in conceptual clarification' in the theory of schemes may have been largely due to the identification between the *points* of a proper scheme X over a discrete valuation *ring A* and the *points* of the tensorial $X \otimes_A K$ over the *field* of fractions K of A.[147] In fact, the usual analytic perspective – busying itself with the observation of points and far from the synthetic reading that would attend to the sections in a sheaf (which do not, in general, proceed from points) – is one that *impedes* scheme theory's emergence, and *obstructs* the *natural change of category* that should have led us to work on rings rather than fields. We thus encounter ways of seeing that are apparently vague and general (analysis *versus* synthesis), but that take on an enormous richness and technical importance that profoundly *affects* the development of precise and well-defined theories. Amid many other examples that we will later review, this *concrete fact* within universal tensions, this *material residue* within ideal dialectics, shows the importance – delineated, instrumental, traceable – of a fundamental analytic/synthetic

one should be able to project the remaining 'Many'-structures that are at stake. What we are dealing with here is neither more nor less than a surprising rebirth of a sort of *mathematical metaphysics* that seeks (and finds) *new archetypes behind the relative*. We will have a chance to discuss this situation at length in part 3.

147 Ibid., 119.

opposition that certain currents of philosophy, following Quine, would rather see disappear.

In a footnote in the EGA, Grothendieck and Dieudonné point out that algebraic geometry, extended to the universe of schemes, should be able to serve 'as a sort of formal model' for analytic geometry (that is to say, for the theory of analytic or holomorphic spaces).[148] Shortly thereafter, Grothendieck will develop his 'Construction Techniques in Analytical Geometry', a remarkable series of notes from the 'seminars', written up in the heat of the moment and *surfacing* in a matter of days.[149] Grothendieck's objective is made explicit at the very beginning of his expositions: to extract ('*dégager*') a general functorial mechanism for the manipulation of modules, applied in particular to the case of complex variables; to extract a 'good formulation' of the problems of modules within the framework of analytic spaces; to interlace projectivity properties with existence theorems within that framework (Teichmüller space); to bring the framework of schemes (algebraic geometry) and the framework of holomorphic manifolds (analytical geometry) closer to one another, in both cases making particular use of the crucial properties of certain nilpotent elements in local rings.[150]

---

148 Ibid., 7.

149 A. Grothendieck, 'Techniques de construction en géométrie analytique I–X', *Séminaire Henri Cartan*, vol. 13 (Paris: Secrétariat Mathématique, 1960–1). They cover about two hundred pages in total.

150 Ibid., exp. 7, 1.

Once again, Grothendieck's generic methods are incarnated in the concrete: the processes of ascent (general functoriality, problematics in abstraction, global frameworks) and descent (projectivity, complex modules, local rings, nilpotence); the dialectic of the One and the Many (the drawing together of frameworks, the specific functoriality of modules, interlacings between projectivity and existence); and, in sum, *the hierarchical structuration of mathematical knowledge in circulating levels*, in which are combined the rich conceptual multiplicity of each object (or morphism), the collection of functors that allows us to 'measure' the differential multiplicity of each level, and the collection of (natural) transformations that let us reintegrate the differential frameworks encountered.

Though we cannot go into excessive technical details here, we may note that Grothendieck's first paper in his 'Techniques of Construction in Analytic Geometry' on its own encapsulates the entire richness of his mathematical thought. The objective is to construct an analytical space of universal representation (Teichmüller space) that classifies all other algebraic curves over analytical spaces;[151] Grothendieck goes on to axiomatically describe the functorial properties[152] that this space should satisfy, and theorematically derives its (global) existence from

---

[151] Ibid., exp. 7, 1.

[152] Ibid., exp. 7, 6. Grothendieck first describes these properties 'vaguely', and only later gives them precise technical descriptions.

the existence of a (local) echeloned hierarchy of adequate fibering functors ('Jacobi functors') that let us control the number of automorphisms of the structures in play (rigid functors).[153] These technical conditions are, in turn, explicated through coverings, group actions, base changes and free operators over the same rigid functors.[154] We thus encounter a genuine mathematics in motion, a relative mathematics that, nevertheless, allows us to encounter certain universal invariants ('Teichmüller space') *in virtue of the sheer variation* of the local mathematical objects, through the subtle hierarchy of mediations that give rise to the global mathematical object.

Grothendieck's legacy in the landscape of contemporary mathematics is becoming increasingly evident as the discipline unfolds and technical advances corroborate many of French mathematician's major intuitions.[155] In the following chapters, we will try to amplify that contemporary spectrum, as we review the works of other mathematicians of the first rank, works that partly continue and partly complement Grothendieck's own, so that later, in the third part of this essay, we may try to reflect – in ontological, epistemological, methodological

---

153 Ibid., exp. 7, 32.

154 Ibid., exp. 7, 18–21 (coverings and groups), 2ff. (bases), 27ff. (functors).

155 This is particularly conspicuous when it comes to his 'Sketch of a Program', with the connections inaugurated there between combinatorics, number theory and functional analysis, which have been extended, to the great surprise of the scientific community, to theoretical physics and cosmology. We will return to this in chapter 6, as we approach Connes and Kontsevich.

and cultural realms – on that extensive spectrum, which is usually ignored in the tracts of mathematical philosophy.

CHAPTER 5

EIDAL MATHEMATICS: SERRE, LANGLANDS, LAWVERE, SHELAH

In the next three chapters we will review other examples of high mathematics in action, in the conviction that advanced mathematics – and, in this case, contemporary mathematics – provides philosophy with *new* problematics and instruments, as we shall see in part 3 of this work. But firstly, in order to present the mathematical landscape more commodiously, we will sort a few of its striking creative contributions into three complementary spectra: *eidal*, *quiddital* and *archeal* mathematics. Of course, these mathematical realms (and the neologisms that denote them) *do not exist in a well-defined way* as such, and must be understood only as expository subterfuges for easing the presentation and to help us get our bearings in a complex terrain.

In fact, behind the central question of phenomenology – How do we *transit* between the Many and the One? – along with its polar subquestions: How do we unify phenomena by means of categories, and how do we multiply the universal in the diverse? – lie crucial *modes of transformation* of knowledge *and* the natural world. Mediations, hierarchizations, concatenations, polarizations, inversions, correlations, and triadifications, for instance,

are series of transformations leading to the partial explication of certain universal categories,[156] in knowledge
(reorganizing the Kantian inheritance as regards the
transit between the *noumenon* and the *phenomenon*) and
the physical world alike.[157] Within that universal *trans-
formism* – present since the very beginnings of Greek
philosophy and, in the field of mathematics, now codified
in the mathematical theory of categories – it has always
been possible to detect a double movement in perpetual
readjustment: an oscillating series of ascents and descents
in the understanding; and a search for invariants behind
those natural oscillations. We will call movements of
ascent eidal (from *eidos* [idea]), movements of descent
quiddital (from *quidditas* [what there is]), and the search
for conceptual invariants in the various forms of transit
archeal (from *arkhê* [principal]).

In its very etymology, eidos involves an interlacing of
seeing (*idein*) and knowing (*oida*). In 'raising' herself from
the world toward the ideas, the observer contemplates an
open landscape from a higher perspective and can 'see'

---

156 This *transformational* operation can be seen in great detail, for example, in the emergence
of the three Peircean categories, as shown in the doctoral thesis of André de Tienne,
L'analytique de représentation chez Peirce. La genèse de la théorie des catégories (Brussels:
Publications des Facultés universitaires Saint-Louis, 1996).

157 This is particularly visible in contemporary physics and biology, which are becoming ever-
more imbued with dynamical considerations, linked to the description-comprehension
of 'diagrams of transit'. As we have indicated throughout this work, mathematics –
straddling the 'pure' understanding and the physical world – incorporates to an even
greater extent, in a visible and conspicuous manner, the study of *general and particular
problematics of transit*.

further. Extended vision thus implies a greater breadth knowledge. In mathematics, the interest in great 'ideas' is no different: they open an immense field of action, by which work programs can be organized, horizons uncluttered, and subspecialists oriented. The ideas, in turn, combine with images (*eidola*), and so often comprise surprising *transfusions of form*. In what follows, we shall see how certain incisive contemporary contributions in mathematics respond, in a technical manner, to sophisticated distillations of form in the conceptual world of mathematical ideas.

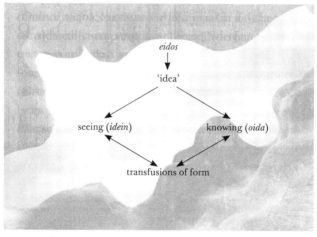

Figure 7. *The Eidal Realm.*

The great interlocutor and proponent of Grothendieck's work, **Jean-Pierre Serre** (France, b. 1926) – a sort of sparring partner for his friend, as we have seen in the previous chapter – may be considered one of the greatest mathematicians of the twentieth century in his own right. Awarded the Fields Medal (1954) and the Abel Prize (2003),[158] Serre has received the mathematical community's highest distinctions, in recognition of his brilliant investigative beginnings,[159] as well as his mature work as an exemplary creator and scientist. Serre's main works cover a very broad mathematical spectrum: the study of the homotopy groups of hyperspheres by means of pathspace fibration (with spectacular calculations, such as the determination of the twelve continuous applications between a 6-dimensional and a 3-dimensional sphere); foundational works at the intersection of algebraic geometry and analytic geometry, based on the emerging theory of sheaves, *FAC*[160] and *GAGA*;[161] the Galois representations associated with formal groups, abelian varieties and modular forms (which, among other conjectures, would interlace Fermat's Theorem with key advances in algebraic

---

158 The Abel Prize citation honors him 'for playing a central role in shaping the modern form of several parts of mathematics, including topology, algebraic geometry and number theory'. Note the importance of *form* in the citation. The casual use of the adjective *modern*, however, turns out to be inadequate according to our own delineations.

159 Serre remains one of the youngest Fields medalists in history.

160 J.-P. Serre, 'Faisceaux algébriques cohérents' (FAC), *Annals of Mathematics* 61, 1955: 197–278.

161 J.-P. Serre, 'Géométrie algébrique et géométrie analytique' (GAGA), *Annales de l'institut Fourier* 6, 1956: 1–42.

geometry and number theory). Serre indicated how his works, despite their *apparent* eclecticism, responded to the same way of observing and transforming problematics, thanks to the use of *transversal* instruments and the emergence of *mixtures* (recall Lautman) in which the One and the Many are technically conjugated:

> I work in several apparently different topics, but in fact they are all related to each other. I do not feel that I am really changing. For instance, in number theory, group theory and algebraic geometry, I use ideas from topology, such as cohomology, sheaves and obstructions.
>
> From that point of view, I especially enjoyed working on *l*-adic representations and modular forms: one needs number theory, algebraic geometry, Lie groups (both real and *l*-adic), *q*-expansions (combinatorial style)... A wonderful mélange.[162]

---

162 M. Raussen, C. Skau, 'Interview with Jean-Pierre Serre', *Notices of the American Mathematical Society* 51, 2004: 211. Regarding the *mode of creation* underlying the emergence of that 'wonderful mélange', it is interesting to point out that Serre speaks of how 'you have some ideas in mind, which you feel should be useful, but you don't know exactly for what they are useful', and of working 'at night (in half sleep)', which 'gives to the mind a much greater concentration, and makes changing topics easier' (in C.T. Chong, Y.K. Leong, 'An Interview with Jean-Pierre Serre', *Mathematical Medley* 13, 1985, http://sms.math.nus.edu.sg/smsmedley/smsmedley.aspx#Vol-13). The combination *of the fuzzy boundary and the great potential for exactitude* is a fascinating theme for mathematical philosophy, which an analytic approach is unable to take up. In chapter 10, we will see how a broader mathematical phenomenology (which incorporates the instruments of Peirce, Merleau-Ponty and Rota, among others) can help us better understand those transits of creation.

Various concrete cases of the 'transfusion of forms' emerge in Serre's work. When, for example, in *GAGA*, Serre establishes an equivalence between algebraic (coherent)[163] sheaves over a projective variety and (coherent) analytic sheaves over the analytic space associated with that variety, an equivalence whereby cohomological groups appear as invariants,[164] he is precisely pinpointing the transit between algebraic and analytic forms, while carefully controlling the osmoses and obstructions in play. When, at the end of the fifties, Serre tests various structures in an effort to produce a 'good' cohomology for varieties defined over finite fields (trying to make some headway on the Weil conjectures in this way), and when the cohomology group with values in a sheaf of Witt vectors shows up[165] (some of the transpositions and obstructions of which will become the very inspiration for Grothendieck's *l*-adic, crystalline and étale cohomologies), Serre once again brings a tremendous labor of precision to bear on the manipulation and transference of forms. In the environment

---

163 Coherence codifies a finite type of property in sheaves. Coherent sheaves come from both analytic geometry (the sheaf of germs of holomorphic functions), and from algebraic geometry (the structural sheaf of a Noetherian scheme). A *common ideal form* thus hides behind coherence. This has to do with a technical *stratum* of unification that allows for a still-greater unity on the higher level of cohomology groups.

164 We are drawing here (and in what follows) on the excellent overview, P. Bayer, 'Jean-Pierre Serre, Medalla Fields', *La Gaceta de la Real Sociedad Matemática Española* 4, 2001: 211–47. This is perhaps the best survey of Serres's work in any language.

165 Witt vectors (1936) are infinite successions of elements in a ring, by means of which sums and products of *p*-adic numbers can be represented in a *natural* manner. They therefore have to do with underlying *forms*, hidden behind certain incomplete representations. Here, one can see how it is that, though *only* certain hierarchies of eidal adequation are under consideration, new mathematics are already emerging.

of eidal mathematics, one can thus catch a glimpse of a complex dialectic that delineates both the movement of concepts/objects (the functorial transit between the algebraic, the geometrical, and the topological) and the relative invariants of form (cohomologies). At stake is a profound mathematical richness – a richness that vanishes and *collapses* if one restricts oneself to thinking in terms of elementary mathematics.

Serre underscores a deep *continuity* in his creative career, beyond certain *apparent* cuts or ruptures:[166] an interlacing of homotopy groups and C-theory (see p. 164, n. 134); a natural osmosis between certain structures from both complex analysis (sheaf cohomologies in the range of functions of various complex variables or complex projective varieties) and algebra (cohomology-sheaves in the range of rational functions or algebraic varieties); a study of algebraic geometry over arbitrary fields (from algebraic closures to finite fields, by way of generalizations of the theory of class fields) in terms of groups and Lie algebras as 'mother' structures, where we find an intersection of the extant contextual information. Indeed, looking over certain contiguities/continuities between the Riemann hypothesis, certain calculations over modular forms and certain calculations of characteristics of discrete subgroups of the linear group, Serre exclaims,

---

166 Chong, Leong, 'An Interview with Jean-Pierre Serre'.

'such problems are not group theory, nor topology, nor number theory: they are just mathematics'.[167] Advanced mathematics thus contemplates a series of sophisticated technical transits over a *continuous conceptual ground*, something that, again, is lost from view when we restrict our perspective to the (essentially discrete) fragment of elementary mathematics.

This continuity of mathematical knowledge has been energetically emphasized in the work of **Robert Langlands** (Canada, b. 1936). The *Langlands program*, in fact, consists of an extensive web of conjectures by which number theory, algebra, and analysis are interrelated in a precise manner, eliminating the official divisions between the subdisciplines. The program first emerges in a long letter (1967) from the young (and unknown) Langlands to the eminent master of the epoch, Weil. The letter turns out to be full of admirable conjectures that approach the world of the complex variable and the world of algebraic extensions *functorially*, by way of group actions. Langlands arrives at these surprising approaches by following the precise elevations of group actions in the *eidal* realm.[168] In effect, Langlands's intuition emerges through a

---

167 Ibid.

168 The extremely meticulous Weil is somewhat irritated by the young Langlands's flights of fancy, and he immediately sends the reckless young man's interminable, handwritten letter to be typeset. The letter to Weil, along with a great deal of additional material by (and on) Langlands, can be found online: http://www.sunsite.ubc.ca/DigitalMathArchive/Langlands/intro.html.

correlative contemplation of two ascending paths: the transit from modular forms (analytic functions of complex variables respecting certain actions of the group $SL_2(R)$) to automorphic forms (analytic functions respecting actions of *Lie groups*), passing through the intermediate actions of the Fuchsian group on Poincaré's modular forms and those of the symplectic group on Siegel's modular forms; the transit in the hierarchy of L-representations[169] of the Galois group $Gal(K^*{:}K)$.[170]

The correspondences between the *forms* of that transit come to express the celebrated *Langlands correspondence*: the automorphic forms associated with the linear group $Gal(n{:}K)$ correspond (functorially) to the *n*-dimensional L-representations of the Galois group $Gal(K^*{:}K)$. The series of *conjectures* thus achieved concretize the continuity of mathematical thought in a remarkable manner.[171]

---

169 The L-series (Dirichlet) appear as analytic objects for representing Riemann's ζ function. The L-functions (Artin, Hecke) are analytic continuations of the L-series that serve to measure the ramification of prime ideals in algebraic extensions. The abstract construction (eidal elevation) of the concept of L-function therefore *integrates* analytic, algebraic and arithmetical considerations.

170 *K* is an arbitrary field and $(K^*{:}K)$ need not be commutative. The commutative case had already been resolved, before Langlands, in *class field theory* (Hilbert, Takagi, Hasse, Herbrand) and, in fact, considerations pertaining to class fields guided a great deal of the Canadian mathematician's intuition.

171 It is worth the trouble to reproduce the beginning of the letter to Weil here: 'While trying to formulate clearly the questions I was asking you before Chern's talk, I was led to two more general questions. Your opinion of these questions would be appreciated. I have not had a chance to think over these questions seriously and I would not ask them except as the continuation of a casual conversation. I hope that you will treat them with the tolerance they require at this stage.' The combination of informality (supposed lack of seriousness, casual conversation, tolerance) and profundity (as the seventeen-page letter is far from lighthearted or frivolous) should remind us of Serre's observations regarding mathematical thought 'in the middle of the night'.

Just as Grothendieck sensed the existence of motifs beneath the various manifestations of cohomologies, Langlands senses the existence of *structural forms of transit* beneath the various natural manifestations of group actions in complex analysis and number theory. In contemporary mathematics, as we shall see in chapter 7, this leads to the recognition of a series of *archetypes*, a series of structures/concepts lying in the depths of the mathematical *continuum*, from which many other partial forms, deriving from the archetype, are extracted through representational 'cuts'.

From the point of view of the global concepts in play, the Langlands correspondence proposes an unexpected equivalence between certain *differentiable* structures associated with an extended modularity (the automorphic forms associated with the linear group) and certain *arithmetical* structures associated with analytic continuations (the L-representations of the Galois group). The profound proximity of the differentiable and the arithmetical, in the restricted context of modular action and analytic continuation, constitutes a truly *major discovery* for contemporary mathematics. In fact, the Langlands program has given impetus to many highly technical results, including various proofs of the correspondence, for every n and for specific cases of the field K under consideration: for fields of formal series over finite fields (Laumon, Rapoport, Stuhler 1993); for p-adic fields (Harris, Taylor 1998); for fields of rational functions over curves defined over finite fields (Lafforgue 2000, work which won him the 2002 Fields Medal).

Nevertheless, the program is currently encountering a *formidable obstruction* in tackling the 'natural' fields of characteristic zero ($\mathbf{Q}, \mathbf{R}, \mathbf{C}$), for whose study, it seems, the indispensable instrumentarium is yet to be constructed. Here, technique encounters one of those paradigmatic conceptual fissures that may be of the greatest interest for the philosophy of mathematics. Now, the leap to the analytic (L-functions) gives us a better understanding of certain fragments of number theory, but then, once this leap has been made, we encounter major obstructions in returning to what *should* be the natural structures of the analytic (fields of characteristic zero). And so we find ourselves faced with new obstructions in transit (obstructions that are carefully maintained, in the scope of the Langlands program, through a sophisticated *theory* of the structural transfusions of form), revealing once again the incessant presence of what, according to Thom, is the 'founding aporia of mathematics': that inherent *contradiction* between the discrete and the continuous that *drives* the discipline. *Fundamentally*, nothing could therefore be further from an understanding of mathematical invention than a philosophical posture that tries to mimic the set-theoretical analytic, and presumes to indulge in such 'antiseptic' procedures as the *elimination* of the inevitable contradictions of doing mathematics or the *reduction* of the continuous/discrete dialectic. In approaching the works of a few of the great contemporary mathematicians, we see how the labor of the 'real' mathematician (in Corfield's sense) points in the *exact* opposite direction: toward the *multiplication* of the dialectic, and the *contamination* of the spaces of mathematical knowledge. At stake is a tremendous leveraging of knowledge, which the philosophy of mathematics should begin to take stock of.

Given an algebraic group[172] G and an L-function, one can construct a new group L_G (the Langlands group) that combines the absolute Galois group over the field underlying G and a complex Lie group associated with L. What is at issue here is a mixture (in the full Lautmanian sense) – a mixture that helps us control the theory of representations of G. Within this framework, some of the underlying transits in the Langlands program correspond to the *functorial fact* (plausible, correct in particular cases but generally undemonstrated) that every morphism $^L G \to {}^L G$ *issues* from a morphism between the associated automorphic forms, which is well behaved in every p-adic stratum of representation, *for almost every p*.[173] We are faced here with

---

172 An algebraic group is an algebraic variety that has, in addition, the structure of a group. Examples of algebraic groups include finite groups, linear groups $GL(n)$, and elliptical curves.

173 See R. Langlands, 'Where Stands Functoriality Today', in T.N. Bailey & A.W. Knapp, *Representation Theory and Automorphic Forms*, Proceedings of the Symposium on Pure Mathematics 61, (Providence: American Mathematical Society, 1997), 457–71.

a recurring situation in advanced mathematics, one that *cannot be contemplated* in mathematical contexts of a lower level of complexity – a situation in which the general forms of knowledge (*universality*, functoriality) *and* the restricted calculations underlying them[174] (*particularity*, diophantine objects) depend on a complex *intermediary* hierarchy that *forcibly* structures both the generic transits (going up) and the specific obstructions (going down).

In an interview, Langlands sums up his perception of mathematics:

> I love great theories, especially in mathematics and its neighboring domains. I fell in love with them, but without really grasping their significance, when I was still a student. [...] What I love is the *romantic side* of mathematics. There are problems, even big problems, that nobody knows how to tackle. And so we try to find a *footpath that leads to the summit*, or that lets us approach it. [...] I love having the impression of standing before a virgin continent. I love the problems whose solutions require unpublished

---

174 We cannot evoke, here, the enormous *concrete richness* of the calculations, but it is enough to imagine that they include the entire, enormous tradition of nineteenth-century and early twentieth-century German arithmetic (Jacobi, Dirichlet, Eisenstein, Kummer, Hilbert, Hecke, Artin, Hasse, etc.).

and unsuspected theories. In other words, I love the mathematics that *make us dream*.[175]

The ascent to the peaks of the eidal thus grants us privileged visions, like the one which Langlands had at age thirty-one, and which he looks back upon with admiration: 'That I had nevertheless arrived somewhere always seemed to me a miracle [...] that I had seen so much in a single blow, I do not think that this will ever cease to astonish me.'[176] His astonishment is sincere and, to a certain extent, tied to the marvels of his discovery (a 'virgin continent'), but one can already observe (though we will make this more precise in chapters 10 and 11) that, in contemporary mathematics, many great creators are able to climb to the summits precisely in virtue of their capacity to *correlatively transit* the world of the eidal, approaching polarities, hierarchies, relative invariants and intermediary structural correspondences in a systematic manner. What we are dealing with here is a *dynamics* of the mathematical world that is completely different from what we can find in elementary mathematics, whose low threshold of complexity does not allow for those aforementioned transformations to emerge. It is also completely different

---

175 S. Durand, 'Robert Langlands. Un explorateur de l'abstrait', *Québec Science* 2000, http://www.crm.umontreal.ca/math2000-1/pub/langlands.html (italics ours). We will come back to the indispensable *romantic side* of mathematics in our final chapters.

176 *Response to the Gold Medal of the French Academy of Sciences* (2000), http://sunsite.ubc. ca/DigitalMathArchive/Langlands/misc/gror.ps .

from what is ordinarily discussed in the analytic philosophy of mathematics, whose dismemberment of mathematical objects, in order to procure a static and ultimate foundation that would sustain them, does not allow us to take stock of their incessant torsions – and so impedes any approach to mathematics' *transitory specificity* (on three levels: phenomenological, ontological and epistemological, as we shall see later on).

Langlands's revealing commentary on the difference between the Taniyama-Shimura Theorem[177] and Fermat's Theorem, shows the importance of a mathematical thinking that is attentive to *eidal generic* structures:

> Fermat's Theorem is an unexpected consequence of another theorem (Taniyama-Shimura-Weil). The latter has to do with a coherent framework, whose beauty in my eyes comes from the fact that it corresponds to an *order* I'm used to seeing in number theory. On the other hand, according to my intuition or imagination, Fermat's Theorem could have turned out to be false without that order being disturbed'.[178]

---

[177] The Taniyama-Shimura conjecture (1955) suggests the equivalence (modulo L-series) of elliptic curves with modular forms. Frey (1985) conjectured that a nontrivial Fermat-style solution $x^n + y^n = z^n$ would yield a nonmodular elliptic curve (a 'Frey curve'). Ribet demonstrated (1986) the Frey conjecture, thus establishing the implication Not (Fermat) $\Longrightarrow$ Not (Taniyama-Shimura), or, equivalently, Taniyama-Shiumura $\Longrightarrow$ Fermat. Wiles (1993–4) demonstrated the Taniyama-Shimura conjecture for semistable elliptic curves (among which appears Fermat's curve), thereby proving Fermat's Theorem. The full proof of Taniyama-Shimura, for all elliptic curves, was finally obtained in 1999 (Breuil, Conrad, Diamond, Taylor).

[178] Durand, 'Robert Langlands...' (italics ours).

General counterweights, pendular oscillations, counterpoints in an abstract order, and hidden aesthetic harmonies therefore guide the *theoretical vision*, whereas the particularity of the concrete case may come to distract it. It is in the *equilibrium* between the broadest abstract universality and the most restricted concrete particularity – that is to say, in the broad register of *mediations* – that mathematical inventiveness emerges with all its force, haloed with '*order and beauty*', luxury, peace and voluptuousness'.[179]

William Lawvere (USA, b. 1937) has been one of the main proponents of a broad hierarchy of mediations within the general horizon of category theory. Since his doctoral dissertation,[180] Lawvere has consistently insisted on thinking differently, and has succeeded in situating himself in a conceptual *underground* where the synthetic and the global take precedence, setting himself at some distance from the usual analytic and local approaches. Lawvere's thought combines various strategies that aim to fully capture the *movement* of mathematical concepts, thereby following the lines of Grothendieck's work. Whether through an 'inversion of the old theoretical program of modeling variation within eternal constancy', or by breaking the '*irresolvable*

---

179 C.Baudelaire, 'L'invitation au voyage,' *Fleurs du mal* (1857).

180 F. W. Lawvere, *Functorial Semantics of Algebraic Theories* (New York: Columbia University Press, 1963). A summary is given in *The Proceedings of the National Academy of Sciences* 50, 1963: 869–72. Lawvere was a student of Eilenberg and Mac Lane, the creators of category theory (Eilenberg, Mac Lane, 'General Theory of Natural Equivalences')

contradiction' that is the 'metaphysical opposition between points and neighborhoods', or by constructing a topos theory that allows for a back-and-forth between constant and variable sets,[181] Lawvere seeks to build a dynamic conception of mathematics, capable of capturing both the physical world's continuous becoming and the continuous enfolding of the relational canopy spread over it.[182] As we shall see, Lawvere's procedure consists in *elevating*, into the eidal, a sophisticated web of weavings and oppositions between concepts, calculi and models, extending them both vertically (hierarchically) and horizontally (antithetically). The synthetic comprehension of that web, 'under the guidance of that form of objective dialectics known as category theory',[183] is a fundamental contribution to contemporary thought – one that is still in the process of transcending the strictly mathematical horizon of its origin.

One of the few great contemporary mathematicians who has dared to interpolate his writings with vague and

---

181 F. W. Lawvere, 'Continuously Variable Sets; Algebraic Geometry = Geometric Logic', in *Proceedings of the Logic Colloquium 1973* (Amsterdam: North-Holland Publishing Co., 1975), 135–56.

182 Lawvere was, first, a student of Truesdell in continuous mechanics; although he later directed himself toward the categorical foundations of mathematics and logic, he has always given preference to the categorical instruments dealing with transit and flux (variable sets and sheaves, beyond the static and the punctual) as being better suited for an understanding of the physical world.

183 F. W. Lawvere, 'Introduction', in *Proceedings of the Halifax Conference on Toposes, Algebraic Geometry and Logic*, (New York: Springer, 1972), 1.

nondisciplinary[184] references as a source of subsequent technical refinements, Lawvere has been able to construct an uneasy equilibrium between the *seer*, who peers into abysses ('gauging to some extent which directions of research are likely to be relevant'[185]), and the *climber*, who secures every step in his ascent (confronting some of the major technical challenges of the age). In a remarkable article on the 'future' of category theory,[186] Lawvere characterizes the theory of categories as

> the first to capture in reproducible form an incessant contradiction in mathematical practice: we must, more than in any other science, hold a given object quite precisely in order to construct, calculate, and deduce; yet we must also constantly transform it into other objects.[187]

Category theory's capacity to *axiomatize*, with great precision,[188] the fundamental *weaving* between static considerations (states, points, objects) and dynamic ones (processes, neighborhoods, morphisms) is one of the deep

---

184 Including up-front and provocative mentions of Engels, Lenin, or Mao.

185 F. W. Lawvere, 'Adjointness in Foundations' in *Dialectica* 23, 1969: 281–96: 281).

186 F. W. Lawvere, 'Some Thoughts on the Future of Category Theory', in *Proceedings of the Como Meeting on Category Theory*, (New York: Springer, 1991), 1–13.

187 Ibid., 1.

188 Lawvere speaks of the 'crystalline philosophical discoveries that still give impetus to our field of study', ibid.

reasons for its success. The theory presents a permanent back-and-forth between the three basic dimensions of the semiotic, emphasizing translations and pragmatic correlations (functorial comparisons, *adjunctions*[189]) over both semantic aspects (canonical classes of models) and syntactic ones (orderings of types). In Lawvere's vision, we find opposed – and, in fact, springing from that very opposition – two classes of categories corresponding to 'Being' and 'Becoming', between which is set to vibrating a 'unity and identity of opposites'[190] that gives rise to remarkable mathematical conjectures in an intermediate terrain between the ascent to the general ('from below, from real space') and the descent to the particular ('from above, by classifying abstract algebra').[191] A 'descending' functor sends a given category to a smaller one and, in two linked movements of oscillation (two adjunctions, see figure 8), two *skeletal* counterpoints appear (in 'positive' and 'negative'[192]) as extreme images of Becoming. The category is *set free* by this back-and-forth: the skeleton surfaces as a static filtration (in Being) *after* its immersion

---

189 A categorical adjunction generalizes a residuation in an ordered set (which is to say, a pair of morphisms f, g, such that $fx \leq y$ iff $x \leq gy$). Adjunction consists in a pair of functors $F, G,$ such that $Mor(FX,Y) \simeq Mor(X,GY)$ (natural isomorphism). Residuations run through all of algebra, and, in particular, the algebra of logic, giving rise to implication and the existential quantifier. In category theory, adjunctions, linked to *free* objects, appear even more ubiquitously.

190 Ibid., 2.

191 Ibid., 12.

192 Ibid., 7.

in a dynamic fluid (in Becoming). The skeletons (positive and negative), together with the descending functor, form a 'unity and identity of opposites', in which what appears as contradictorily fused on one *level* can be separated and distinguished on another. The descent to the abyss happens to be perfectly controlled by a hierarchical strategy – levels and contexts, which is to say, functors and categories – in a way that is quite close to Peircean pragmatism.

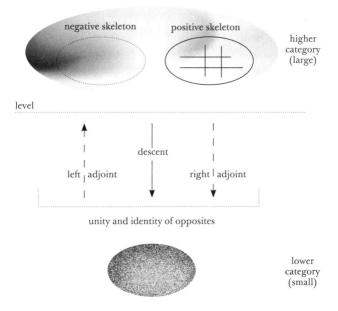

Figure 8. *'Unity and identity of opposites': descents, levels, and skeletons, according to Lawvere.*

Lawvere's examples cover a multitude of modern and contemporary mathematical fields: general and algebraic topology; algebraic and differential geometry; abstract algebraic structures – categories, lattices, rings – nonclassical logics; functional analysis; mathematical physics; etc. Within the framework of that collection of concrete situations, Lawvere nevertheless performs an incessant eidal exercise: knowledge *emerges* in the wake of a meticulous strategy of descent and ascent, combined with a dialectic of contrapositions. In Lawvere's words, 'This explicit use of the unity and cohesiveness of mathematics sparks the many particular processes whereby ignorance becomes knowledge'.[193] The processes of weaving, the progressive freeing of objects,[194] the contraposition of opposite skeletons allows us to stand at a distance, decant the objects, and look on with *other eyes*.

The union between the static and the dynamic anticipated by Novalis is realized with the greatest originality in the theory of categories. A *natural* combinatorics of levels allows a single object to be represented as simultaneously fixed (in a given category) and variable (through

---

193 Ibid., 2.

194 A *free object* in a category has a precise technical definition, but can be seen as a universal object – disincarnate and skeletal – with astonishing projective ductility; it has the capacity to project, in a unique manner, its *entire* formal structure onto any other object in the category having *no more* than a similar basis. Free objects allow for a remarkable formal passage from the part to the whole, in mathematical situations involving rich possibilities of transit (which are not always given, since not every category possesses free objects). Within this framework, an *adjunction* can be seen as a generic, uniform and cohesive process of constructing free objects.

its functorial transformations). Schlegel's aphorism that links universality and transformation – 'the life of the Universal Spirit is an unbroken chain of inner revolutions',[195] yet another form of the 'Great Chain of Being' and the infinite 'Tree of Knowledge' studied by Lovejoy – is, in category theory, incarnated with precise sophistication. The universals of the theory are always dynamic universals, never rigidified in a fixed Absolute. The entire set of categorical instruments – composition, morphisms, natural transformations, sketches, limits, adjunctions, sheaves, schemes, etc. – is converted into an enormously powerful technical arsenal that unexpectedly revitalizes the romantic dialectic between Being and Becoming. We are not far here from the 'romantic side' of mathematics divined by Langlands.

Within this very general dynamic framework, Lawvere's work manifests *concrete distillations of forms* in multiple manners. An admirably simple fixed-point argument in closed cartesian categories lets us interlace Cantor's Theorem $card(X) < card(\wp X)$, Gödel's Incompleteness Theorem, Tarski's satisfiability theory and the occurrence of fixed points in complete lattices.[196] A subtle understanding of certain exactness properties in topoi allows us to interpret

---

195 F. Schlegel, *Philosophical Fragments*, tr. P. Firchow (Minneapolis: University of Minnesota Press, 1991), 93 (fragment 451).

196 F. W. Lawvere, 'Diagonal Arguments and Cartesian Closed Categories', in *Lecture Notes in Mathematics* 92 (New York: Springer, 1969), 134–45.

quantifiers as adjoints, describe the logical behavior of sheaves, and launch a program for the geometricization of logic in which, to everyone's surprise, intuitionism takes a central place.[197] An elementary axiomatization of topoi (with regard to the subobject classifier) allows us to reinterpret topologies as modal operators, abstract the properties of Grothendieck topologies by means of closure operators j over the subobject classifier, and define an abstract notion of sheaf with respect to those 'topologies' j.[198] These are all examples of how the various subfields of mathematics tend to *contaminate one another*, and how certain formal subspecializations *transit simultaneously* through logic, geometry and algebra. Indeed, we find ourselves before a genuine ontic/epistemic break, emerging in modern mathematics and fully defined in contemporary mathematics, in which the old subdivisions of the discipline tend to *disappear*.

---

197 F. W. Lawvere, 'Quantifiers and Sheaves', in *Actes du Congrès international des mathématiciens 1970*, (Paris: Gauthier-Villars, 1971) 329–34. Lawvere explicitly indicates that 'in a sense, logic is a special case of geometry' (329). This is the announcement of a growing geometricization of logic, some of the manifestations of which we will review in chapter 7 (Freyd, Zilber), and which constitutes a renewed *geometricization of contemporary mathematics*, as we have indicated in chapter 2. It may be noted that the Nice Congress, where Lawvere proposed his program, is also where Grothendieck announced his premature abandonment of the mathematical world. Though it could not be clearly seen at the time, the Frenchman's ideas were to undergo a dramatic reincarnation at the hands of the North American.

198 F. W. Lawvere, 'Introduction', in *Lecture Notes in Mathematics 274*, (New York: Springer, 1972), 1–12. Lawvere and Tierney axiomatized the topoi (coming from the Grothendieck school) in elementary terms, which is to say, without requiring conditions of infinity or choice. A subobject classifier generalizes the idea of power set, and allows the *sub*constructions of ordinary set-theoretical mathematics to be governed through the universal behavior of the morphisms in play.

The work of **Saharon Shelah** (Israel, b. 1945) offers new and profound technical arguments for understanding a powerfully *stratified* mathematics, imbued with multiple *transversal* tensions, and attends to the study of that stratification's *structural limitations*, both in its horizontal levels and along its vertical skeleton. Lying far beyond the Gödel's Incompleteness Theorem (regarding the *deductive* limitations of theories whose threshold of complexity meets or exceeds that of Peano arithmetic), Shelah's *nonstructure theorems* (1980–90) uncover the *semantic* limitations of natural classes of models in advanced mathematics.[199] Shelah's results reveal an unexpected *polarization* in the study of classes of models for a classical theory T. His Main Gap Theorem shows that the number of nonisomorphic models (of a given size) for T faces an abrupt alternative: *either* it literally explodes, reaching the maximum possible number of models, *or else* it turns out to be perfectly controllable. There is no place for semicontrol or semi-explosion: either the class of models for T cannot count on having any structure at all (all of the possible models are given, everything that can *possibly* exist also exists *in actuality*); or else the class can be fully structured (all that actually exists also exists in a 'coordinated' form, according to subtle scales

---

199 See the works collected in the monumental (739-page) text, S. Shelah, *Classification Theory and the Number of Nonisomorphic Models*, (Amsterdam: North-Holland, 1990).

of invariants). Indeed, the impetus behind Shelah's most original ideas is ultimately the acquisition of a *general theory of dimension* (subtle invariants for the structured case). His 'theory of excellence', in particular (the buttress of the difficult part of the Main Gap proof, which took ten years to complete) requires a series of 'algebraic' interactions in arbitrarily high finite dimensions, which by far transcends the usual 2-dimensional interactions that show up in the independence proofs of traditional algebraic geometry.[200] We thus find ourselves faced with yet another situation in advanced mathematics where a leap in complexity gives rise to new mathematics, which are *not at all reflected* in the lower strata.

After detecting the *generic* presence of the gap in the set-theoretical universe, the tremendous labor of Shelah and his team[201] was devoted to describing many observable *concrete* conditions, in order to be able to detect whether an arbitrary theory can be *classified* as structure

---

200 Thanks to Andrés Villaveces for these clarifications. According to Villaveces (personal communication), 'There is a quantity of structures "waiting to be discovered", in geometry and in algebra, that *demand* that the algebraic interaction accounts for all those high-dimensional diagrams. [...] Already in group theory they're beginning to make 3-dimensional amalgams. They are very difficult and correspond to group-theoretic properties that are truly more profound than most of the traditional ones. [...] It is sort of as if, in geometry, we had been working, up to now, with a "2-dimensional projection" of phenomena that would have seemed more natural if we had contemplated them in their true dimension.'

201 Shelah's charisma has given rise to a genuine logic workshop, distributed over numerous countries. The work of Shelah and his collaborators is nearing a thousand articles, almost altogether implausible in the mathematical world. In virtue of the profundity of his general ideas, his technical virtuosity, the tenacity of his daily work, and his influence in the community, Shelah can easily be seen as the greatest logician of the twenty-first century.

or nonstructure. The classes of models for such a theory range, in principle, between two extremes: proximity to a Morley-style categoricity theorem, where the models turn out to be isomorphic by strata,[202] or else freedom from any structural restriction. A first dichotomy for classifying classes of models distinguishes between *stable* and unstable theories. The deep mathematical meaning of stable theories comes from the structure of the complex numbers **C**, with *addition and multiplication*, and from the algebraic geometry that can be undertaken within it. The current notions of dimension and algebraicity extend to stable theories and can be used as natural logico-algebraic invariants for 'coordinating' the models of those theories. By the same token, unstable theories are theories in which certain generic orders can be defined;[203] in that case the classes of models tend to disintegrate and their diversity explodes. One example that is now being thoroughly

---

202 The Morley Theorem (1965) asserts that a denumerable first-order theory that is categorical in a nondenumerable cardinal κ (that is to say, such that all of its models of size κ are isomorphic) is likewise categorical in *every* cardinal greater than κ. This is the strongest possible result of the 'collapse' of the infinite for first-order theories (collapse by strata), since, by contrast, from one stratum to another, far from collapsing, the models are multiplied, due to the properties of first-order logic (compactness, Löwenheim-Skolem).

203 We face a situation contrary to that of **C**, in which we cannot define an order congruent with the operations. The fact that complex numbers are not an ordered field, for many decades seen as an important limitation in the architectonic construction of sets of numbers, has today come to be seen as a strength (as a reason for stability). The *pendularity* of the mathematical understanding is evident here. No description of the *ontic richness* of **C** should be allowed to neglect this fundamental oscillation. Nevertheless, not only has this pendular movement gone unstudied in the analytic tradition; its existence has not even been registered! This is one example, among many others in advanced mathematics, that *forces* us to change our philosophical perspective, if we are *really* to be in a position to accept the advances of the discipline.

studied (and that we will come back to in chapter 7, when dealing with Zilber) is the structure of the complex numbers, with addition, multiplication and *exponentiation* added on; the complex exponential in fact introduces a sophisticated hierarchy of analytic submodels that remains beyond the control of first-order logic, and the theory turns out to be profoundly unstable.[204]

Beyond the stable/unstable dichotomy, Shelah's program broaches the problematic of describing and studying many other *dividing lines* by which the Main Gap may be refined, with important *mathematical* (and not just logical) content on each side of the division. One robust dividing line is the dichotomy: superstable + non-DOP + non-OTOP / not superstable or DOP or OTOP, with powerful structure theorems on the superstable + non-DOP + non-OTOP side, where the models can be analyzed by means of *trees* of *countable* models.[205] We then see that a general *eidal polarity* can come to be incarnated in multiple concrete polarities.

---

204 For complements and clarifications, see the existent overview, A. Villaveces, 'La tensión entre teoría de modelos y análisis matemático: estabilidad y la exponencial compleja', *Boletín de Matemáticas* Nueva Serie XI, 2004: 95–108.

205 Omitting types order property (OTOP) indicates that a certain order that is not definable in first-order logic becomes so through the omission of types (in the logic $L_{\omega_1\omega}$ of the greatest expressive power). Dimensional order property (DOP), is another form under which an order that is *hidden* from the eyes of first-order logic may be expressed. Thanks to Andrés Villaveces for this information.

In a prospectus on the future of set theory,[206] Shelah reflects that the main sources of interest in the theory's development have their roots in its beauty (earning it nine points), its generality (six points), its concrete proofs (five points), and its wealth of internal developments (four points).[207] He shores up this polemical vision by adding, 'My feeling, in an overstated form, is that beauty is for eternity, while philosophical value follows fashion'.[208] For Shelah, beauty is rooted in 'a structure in which definitions, theorems and proofs each play their part in the harmony'.[209] *Even though* many of Shelah's major theorems exhibit, analyze and synthesize the inharmonious and nonstructured behavior of certain classes of models, it should be observed that, *in its entire conception*, there nevertheless exists a pendular counterpoint between the structured and the lack of structure, and that oscillation is *in itself* profoundly harmonious. Once again, transfusions and obstructions of forms, with global equilibria and incessant local tensions, govern the main outlines of a decisive body of work in contemporary

206 S. Shelah, 'The Future of Set Theory' (2002), http://arxiv.org/pdf/math.LO/0211397.pdf.

207 Shelah completes the list with other minor sources of interest (in descending order): applications, history, 'sport', foundations, philosophy.

208 Ibid., 2. Note that the *major exponent of set theory* neglects its supposed philosophical value. This fact should cause philosophers who see the philosophy of mathematics only in the philosophy of set theory to seriously question their perspective.

209 Ibid. As examples of beauty, Shelah proposes Galois theory (and 'more exactly what is in the book of Birkhoff-Mac Lane') and the Morley Theorem (with its proof). Observe how a great mathematician insists on the *form* of proofs and on their *exposition*; once again the fundamental pertinence of *style* in mathematics enters the scene.

mathematics. As often happens with great creative mathematicians, advances in one direction of thought find a counterpoint in unexpected advances in the *opposite* direction. After becoming a specialist in *relative consistency* results in set theory,[210] and above all after his very difficult demonstration of the independence of the Whitehead problem,[211] Shelah *turns* toward a new understanding of approximations in the infinite, with an ambitious program of cardinal arithmetic in which he proposes to *redesign adequate invariants* for the operations.[212] Given that cardinal exponentiation effectively turns out to be a great *obstruction* to cardinal arithmetic – due to the 'wild' behavior of 2^κ for an infinite cardinal κ (Cohen's 1963 independence results) – Shelah proposes to seek out a *robust alternative skeleton* for infinitary operations. Shelah finds the supports of that skeleton in his PCF theory (an acronym for '*p*ossible *c*ofinalities'), in which

---

210 Given a statement φ and a subtheory T of the ZF set theory, φ is *relatively* consistent with T if $Con(T) \Rightarrow Con(T+\varphi)$, where $Con(\Sigma)$ means that the theory is consistent, which is to say, that one cannot deduce a contradiction from Σ. φ is *independent* from T if both φ and $\neg\varphi$ are relatively consistent with T. Gödel (1938) began this line of study, with the relative consistency of the continuum hypothesis with respect to ZF. By other routes, relative strategies later came to be converted into one another, as we have seen, in one of Grothendieck's major predominant tendencies.

211 The Whitehead problem (1950) aimed to characterize a *free* abelian group A by means of a condition on its contextual behavior (residual A condition: for every morphism g over the group A, with nucleus Z, there exists a section s such that $gs = id_A$). Shelah demonstrated that, for abelian groups, the conjecture θ (A is residual $\Rightarrow A$ is free) is *independent* of ZF set theory, since, on the one hand, $V = L$ implies θ (hence $Con(ZF+\theta)$), and, on the other hand, $MA+\neg HC$ implies $\neg\theta$ (hence $Con(ZF+\neg\theta)$).

212 S. Shelah, *Proper and Improper Forcing*, (Amsterdam: North-Holland, 1992); S. Shelah, *Cardinal Arithmetic* (Oxford: Oxford University Press, 1994).

he introduces a web of *tame algebraic controls*[213] for cardinal cofinalities,[214] and discovers that, beneath the erratic and chaotic behavior of exponentiation, lies a *regular* behavior of certain reduced products by means of which the upper reaches of the cardinals can be approximated. What we encounter here, again, is the construction of an intermediate hierarchy that allows for the relative adequation of transit and the discovery of its proper invariants.

A sort of correlation (PCF/cardinals $\equiv$ algebraic topology/topology) issues from Shelah's works. In fact, the search for a robust skeleton (cofinalities) and a tame algebraic calculus (reduced products) in the theory of singular cardinals corresponds to the idea of seeking natural algebraic invariants (homotopies, homologies) for topology. This closes one of the circles we entered in this chapter, beginning with Serre's works on the homotopy of spheres.[215] If perpetual transfusions of form have been the

---

213 The strategy of a 'tame' mathematics, which neglects certain *singularities* (such as those artificial counterexamples of general topology, based on the axiom of choice, or such as cardinal exponentiation), goes back to Grothendieck (see p. 147 n. 102).

214 The *cofinality* $co(\kappa)$ of a cardinal κ is defined as the minimal cardinal of the cofinal subsets in (the order of) κ. A cardinal is *regular* if it is equal to its cofinality (example: $\aleph_{a+1}$) and *singular* otherwise (example: $\aleph_\omega$). PCF theory helps to control the subsets of a singular cardinal through cofinalities, something which cannot be done through exponentials.

215 The circle is closed even tighter if we observe (Villaveces, personal communication) the parallel PCF/cardinals $\equiv$ schemes/varieties. In effect, cardinal arithmetic is *localized* in PCF, by controlling cofinalities around fixed cardinals and then 'gluing' the information together – in a manner similar to that by which schemes help to localize the arithmetic of varieties, by controlling local rings over primes and then 'gluing' the information together. In the third part of this essay, we will come back to the *crucial importance of sheaves* – underlying processes of localization and gluing, or, more generally, of differentiation and reintegration – in order to try to capture the passage from modern to contemporary mathematics in an *intrinsic* fashion (and not merely diachronically, in the environment 1940–50).

fundamental *motif* of this chapter, we have also been able to contemplate a rich multiplicity of concrete *modulations* in which this motif is diversely incarnated. The majority of these transits have taken place in the world of the eidal, in the vast space of the mathematical imaginary.

In the next chapter we will see how, in turn, those ascents of mathematical inventiveness succeed in *descending* once again into the physical world, in the most unsettling ways possible. In fact, in the last thirty years, mathematical physics has been impregnated with an extraordinary host of abstract methods from high mathematics (to a large extent through a rejuvenated perspective on Grothendieck's work, thanks to the Russian school), the technical consequences of which we are just now beginning to glimpse, and the philosophical consequences of which may turn out to be utterly explosive.

CHAPTER 6

QUIDDITAL MATHEMATICS: ATIYAH, LAX, CONNES, KONTSEVICH

Since its very beginnings, mathematics has been very close to physics. The observation of natural phenomena has sought to avail itself of the mathematical apparatus at every moment of history. Mathematics, which has always been described as the *universal language* of the sciences, had, at the beginnings of modern mathematics (thanks to Riemann's spectacular revolution) come to understand itself as a sort of *structural machinery* for the sciences. In Riemann's vision, far from being reduced to a mere language, an expedient that would serve only to display what *other* sciences had discovered, mathematics would in fact be the discipline that allows us to codify the deep structures underlying the natural world. The situation was complicated even further in the last half of the twentieth century, with some of the most formidable advances in contemporary mathematics. As we shall see, certain arithmetico-combinatorial structures (the Grothendieck-Teichmüller group) may come to govern certain correlations between the universal constants of physics (the speed of light, the Planck constant, the gravitational constant), while, conversely, certain mathematical theories originating in quantum mechanics (noncommutative geometry) may help to resolve difficult problems in arithmetic (the Riemann hypothesis). We are dealing here with *absolutely unanticipated* results, which bring together the most abstract mathematical inventions and the most concrete physical universe. The problems that these *new* observations pose for the ontology of mathematical objects are enormous: Where do these objects 'live' – in arithmetic *or* in the physical world? Can this alternative really be contemplated? Can we instead elaborate a transitory, nonbipolar ontology? We will be sure to tackle them in part 2 of this work.

In this chapter, we will use the neologism *quiddital* (from *quidditas* [what there is]) to designate the process of *descent* from the highly abstract constructions of contemporary mathematics and their application to the physical world (what there is). This Being is subdivided by a tense contraposition between essence (*ousia*) and existence (*huparxis*), and by a *counterpointing*[216] *of transfusions of the real* that should remind us of the mathematical dialectic between essence and existence studied by Lautman.

---

216 A fundamental neologism introduced by Fernando Ortiz (1940), in his understanding of Latin America. See his *Contrapunteo cubano del tabaco y el azúcar* (Havana: Jesús Montero, 1940) [Tr. H. de Onís as *Cuban Counterpoint: Tobacco and Sugar* (New York: Alfred A. Knopf, 1947)].

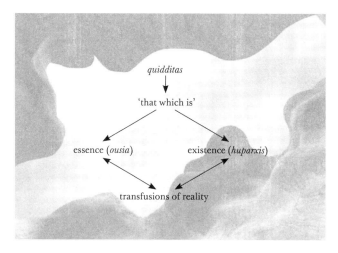

Figure 9. *The Quiddital Realm.*

The works of Sir **Michael Atiyah** (England, b. 1929) have given a decisive impulse to the possible application of contemporary mathematics' sophisticated instruments to the understanding of associated physical phenomena. Awarded the Fields Medal (1966)[217] and the Abel Prize (2004), Atiyah is best known for his famous Index Theorem (in collaboration with Singer, 1963),[218] a very deep result that can be considered one of the major theorems of

---

217 The 1966 Fields medalists were Atiyah, Cohen, Grothendieck and Smale: an impressive mathematical revolution in full bloom!

218 M. Atiyah & I. Singer, 'The Index of Elliptic Operators on Compact Manifolds', *Bulletin of the American Mathematical Society* 69, 1963: 322–433. Further developed in M. Atiyah & I. Singer, 'The Index of Elliptic Operators I-V', *Annals of Mathematics* 87, 1968: 484–604 and 93, 1971: 119–49.

the twentieth century. The theorem effectively combines: a statement of great simplicity and universality (a precise interlacing of transferences and obstructions in the domain of elliptic equations); a very diverse collection of proofs, drawn from apparently contrasting realms of mathematics (K-theory, Riemann-Roch theory, cobordism, heat equation, etc.); a remarkable radiation into a very broad mathematical spectrum (differential equations, mathematical physics, functional analysis, topology, complex analysis, algebraic geometry, etc.).

In rough, conceptual terms, the Index Theorem states that the *balance* between transits and obstructions in certain changes in nature is completely characterized by the *geometry* of the environment where this change takes place. In more precise terms, given an elliptic differential operator ('change'), the index ('balance') – defined as the number of solutions ('transits') *minus* the number of restrictions ('obstructions') for the operator – is completely determined by adequate topological invariants ('geometry of the environment'). In still more precise terms, if we give ourselves an elliptical differential operator[219] D, and if we define an *analytic index* for D with $ind_{analyt}(D) = \dim Ker(D) - \dim coKer(D)$ (kernel $Ker(D)$: 'solutions' – that is,

---

219 Elliptic operators (whose coefficients in higher-order partial derivatives satisfy a suitable condition of *positivity*) appear ubiquitously in mathematics: the Laplace operator ($y_1^2 + \ldots y_n^2$) associated with the heat equation; the Toeplitz operator (given a continuous f, take the holomorphic part in the multiplication of f by a holomorphic), associated with the Cauchy-Riemann equations; the Fredholm operator (derivation in the tangent bundles of a manifold), associated with ellipticity equations in fibers.

harmonic functions; cokernel $coKer(D)$: 'obstructions' – that is, restrictions in nonhomogeneous equations of the type $Df=g$), the Index Theorem asserts that the analytic index can be characterized by means of a *topological index* $ind_{top}(D)$ linked to purely cohomological invariants[220] of the operator's geometrical environment. A fact of the greatest interest that follows from this theorem is that we can observe how the solutions and obstructions, which are *separately* completely *unstable* (owing to the great local variation in the differential equations), nevertheless turn out to be *stable in their difference* (global unification of the indices). We will come back to the profound philosophical interest in results like this, which, again, do not exist in elementary mathematics. But we can already sense the richness of a philosophical approach that really takes seriously the *dialectic of contrapositions of mathematical flows*, *demonstratively* incarnated in the Index Theorem.

The Index Theorem affords a striking quiddital transit between analysis and topology, with all sorts of applications, since elliptic equations serve to model many situations in mathematical physics. But even more astonishing is that transit's *eidal support*, rooted in the *hidden*

---

220 This is a sophisticated invariant that involves, among other constructions, the Thom isomorphism between the homology groups of a manifold and their cotangent space (modulo boundary), the Chern characters coming from K-theory, and the Todd classes of a manifold. Without being able to enter into technical details, we can see how eminently abstract concepts thus appear, typical of the *transfusions of form* that we were able to contemplate in the previous chapter, and which are here applied to the transfusions of the real.

algebraic-geometrical foundations underlying the theorem's proof. The genesis of the Index Theorem is revealing in this sense. The Riemann-Roch Theorem (1850) proposes to algebraically (dimensionally) control the space of meromorphic functions associated with a curve *through* a topological control (the genus) of the Riemann surfaces associated with the curve. It thus presents a first *great translation* of mathematical concepts, which yields the *general problematic* of how the solutions to certain linear systems of algebraic parameters can be controlled *through* appropriate topological invariants. The generalizations of the Riemann-Roch Theorem would later turn out to be legion, and they are situated at the (diachronic) frontier between modern and contemporary mathematics: Schmidt (1929, for algebraic curves); Cartan-Serre-Hirzebruch (1950–56, for *one* system of sheaves); Grothendieck (1957, for *all* systems of sheaves with algebraic parameters: K-theory);[221] Hirzebruch-Atiyah (1958, for all sheaves with continuous parameters: topological K-theory). As a consequence of these advances in the *generic problematic* of transit between

---

221 Grothendieck's K-theory (K for *Klassen*: the study of classes in their *totality*) aims to study algebraic classes of sheaves, and shows how to pass from natural structures of monoids of sheaves to certain *rings* of sheaves (by way of formal inversions in the sheaf's vectorial fibers). Out of K-theory emerges the famous conjecture of Serre (1959) – every finitely generated projective module over $K(X)$ is free – which was settled in the affirmative in 1976, by Quillen and Suslin. (Quillen's 1978 Fields Medal was, in part, due to that proof.) Compare the fortune of Serre's conjecture with that of Shelah's work on the undecidability of the Whitehead conjecture: we should be *struck* by the complexity of a universe in which such apparently similar statements nevertheless find themselves so profoundly *demarcated*. It is a (romantic) *abyss* into which the philosophy of mathematics must enter.

the algebraic and the topological on the basis of complex analysis (Riemann-Roch), Gelfand, in 1960, proposes a generic statement concerning the homotopy invariance of the index. The rupture finally emerges in 1963, when, working with elliptic differential equations instead of linear systems with parameters, and considering algebraic functions as holomorphic functions satisfying ellipticity (Cauchy-Riemann equations), Atiyah and Singer introduce the radical shift in perspective that leads them to combine the statement of the Index Theorem (à la Gelfand) with the entire instrumentarium of concepts inherited from Riemann-Roch (à la Grothendieck).

In Cartan's presentation on Atiyah's work,[222] which he gave when the latter was awarded the Fields Medal, and in Atiyah's later retrospective,[223] both mathematicians underscore the importance of the analytic index ind_{analyt} being *stable under perturbations* and how it is that we can, thus, reasonably hope for a topological formula ind_{top} in purely geometrical terms. Since K-theory (whether algebraic or topological) provides the precise mathematical instrumentarium for capturing extensions of morphisms between (algebraic or topological) structures linked to perturbations, Atiyah remarks that 'K-theory is just the right tool to study the general index problem' and that 'in

---

222 H. Cartan, 'L'oeuvre de Michael F. Atiyah' (1966), in M. Atiyah, D. Iagolnitzer, eds., *Fields Medallists' Lectures* (New Jersey: World Scientific, 2003), 113–18.

223 Atiyah, 'The Index of Elliptic Operators' (1973), ibid., 123–5.

fact, the deeper one digs, the more one finds that K-theory and index theory are one and the same subject!'[224] The horizon we are facing here is similar to the one we passed through in Grothendieck's work: incorporating a transit between objects (variations, perturbations) so as to *then* proceed to determine certain partial stabilities (invariants) beneath the transit. Note that this general strategy, in the most diverse subfields of mathematics, gives rise to *remarkable concrete forms of original knowledge*, forms that go completely unnoticed in static realms where movement is lacking (for example, in the realms of elementary mathematics, where levels of complexity are low).

Atiyah remarks that 'any good theorem should have several proofs, the more the better'.[225] Such is the case with the Index Theorem, which, in virtue of its very centrality, benefits from techniques from many domains; each proof and each point of view goes on to amplify the 'freedom', 'variety' and 'flexibility' of the mathematician.[226] As Atiyah describes it, mathematics, in such various decantations, 'is always continuous, linked to its history, to the past', and it 'has a unity', which nevertheless should not become 'a straitjacket. The center of gravity may change with time. It is not necessarily a fixed rigid object in that sense;

224 Ibid., 133, 134.

225 M. Raussen and C. Skau, 'Interview with Michael Atiyah and Isadore Singer', *Notices of the American Mathematical Society* 52, 2005: 225.

226 Ibid., 226, 227.

I think it should develop and grow'.[227] As we have come to see, in advanced mathematics (whether modern or contemporary) the *transit* of objects is vital, and the variation of the discipline's centers of gravity seems to be no less inevitable. In the *dynamics* of mathematical activity, Atiyah recalls that 'a theorem is never arrived at the way logical thought would have you believe', that everything is 'much more accidental', and that 'discoveries never happen as neatly' as posterity thinks.[228] Moreover, no reduction of mathematics to a series of logical analyses can be anything but a (philosophically unacceptable) impoverishment of the discipline.

The richness of a quiddital mathematics, in profound proximity to reality,[229] is reinforced by the works of **Peter Lax** (Hungary/USA, b. 1926) in two privileged realms where the variability of the real is modeled with precision: differential equations and the theory of computation. The approach to the quidditas in Lax consists in a sort of *pendular oscillation*, the inverse of the movement that we have observed in Atiyah. The latter produces a *descent from the eidal into the quiddital* – from Atiyah's technical mastery in algebraic topology we pass to its subsequent application to the Index Theorem. Inversely, in Lax, a very concrete

---

227 Ibid., 225, 230.

228 Ibid., 225.

229 According to Atiyah, 'almost all of mathematics originally arose from external reality'. Ibid., 228.

originary pragmatics in the quidditas leads to an ascent into the eidal, so as to be able to then *project itself* onto the fragment of reality with which it began. If, indeed, the mathematical 'heart' of the effort to capture the physical world is to be found in *partial differential equations*, and if an adequate 'tomography' of that heart is to be found in the *computational calculations of those equations' solutions*, then the knowledge lying at the intersection of those two fields – precisely Lax's speciality – will help to describe something of a 'real ground' of mathematics.

The citation of the 2005 Abel Prize awarded to Lax underscores 'his groundbreaking contributions to the theory and application of differential equations and to the computation of their solutions'.[230] In Lax's own words, these contributions can be sorted into four essential subareas:[231] integrable nonlinear hyperbolic systems;[232] shock waves;[233] the Lax-Phillips semigroup in scattering

---

230 Portal of the 2005 Abel Prize, http://www.abelprisen.no/en/prisvinnere/2005/index. html.

231 C. Dreifus, 'From Budapest to Los Alamos. A Life in Mathematics', *The New York Times*, March 29, 2005, http://www.nytimes.com/2005/03/29/science/29conv.html.

232 Integrable systems are systems of differential equations for which there exists a well-defined collection of 'conserved quantities' (codified in the *spectrum* of the differential operator), by means of which one can achieve a complete knowledge of the system's solutions. What we are dealing with here is yet another incarnation of one of the central problematics of modern and contemporary mathematics: the study of transits (here, partial differential equations) and invariants (here, conserved quantities) in the realms of exact thought.

233 Shock waves are perturbations that propagate energy in a given (usually fluid or electromagnetic) medium, and that are characterized by an abrupt discontinuity in their initial conditions. Supersonic waves constitute a paradigmatic case.

theory;[234] solutions of dispersive systems when dispersion approaches zero. We can reinforce the *pendularity* between Atiyah and Lax by observing how both mathematicians cover *complementary* fragments in the universe of differential equations: elliptic operators (Atiyah; paradigm: heat equation) and hyperbolic operators (Lax; paradigm: wave equation). A hyperbolic system can be associated, in a natural manner, with 'conservation laws', by means of suitably integrating the flows inhering in the system's *evolution*. In this realm, Lax works according to various alternative strategies: a general theory and structure (transformations that preserve the spectrum of a hyperbolic differential operator); particular theories and structures (entropy conditions in the case of shock waves, the 'Lax pair' for comprehending solitons in the KdV equation),[235]

---

234 Scattering studies forms of deviation in radiation that are due to certain deficiencies of uniformity in the medium in which the radiation is propagated. Many forms of dispersion in the physics of elementary particles (including X-rays) are paradigms of scattering. Scattering techniques, in turn, allow radar photos to be interpreted.

235 The KdV equation (after Korteweg-De Vries, 1895: $u_t + u_{xxx} + 6uu_x = 0$) is one of the best-known examples of a *nonlinear* hyperbolic equation. The equation models the behavior of waves in a fluid (a liquid surface, for example) and has been particularly helpful for applications in naval architecture and for the study of tides. The KdV equation gives rise to a completely integrable system, and its solutions (*solitons*) behave well, since they can be described as *solitary* waves that uniformly displace themselves in the medium, repeating the same pattern of propagation (Kruskal & Zabusky, 1965). Knowledge of those solitons can be *linearized* by means of methods that are inverse to those of scattering, and the 'Lax pair' allows us to discover that inversion by means of suitable noncommutative linear operators. We see here how a very concrete quiddital situation (waves in water, the KdV equation) gives rise to an entire, subsequent eidal architecture (integrable system, solitons, the Lax pair), which then subsequently returns in the quidditas. But the richness of mathematical transits is not restricted to just one direction. Indeed, Kontsevich (1992) has succeeded in demonstrating a conjecture of Witten's, according to which the generative function of the intersection numbers of spaces over algebraic curves (*moduli*) satisfies the KdV equation. In this manner, some of the most abstract constructions in mathematics are governed by an

specific computational calculations (the 'proximity' of a general system to an integrable system[236]).

Lax emphasizes the importance of 'looking at problems in the large and in the small', of 'combin[ing] both aspects' and then benefiting from their combinatorial 'strength'.[237] What we are dealing with here is an oscillating conceptual equilibrium that Lax himself takes his *style* to reflect, a style that seeks a certain elegance, understood as revelation, simplicity and equilibrium between the abstract and the concrete – an elegance that ought to be reflected in the diversity of proofs an important mathematical theorem deserves. In this manner, the richness of mathematical thought, according to a practitioner of the first rank of the discipline, is rooted in the *multiple transits* of proof, and not in the *fixed* statement being demonstrated. We therefore, once again, see how any logical reduction (or tautological reduction, in the style of the early Wittgenstein) of a statement with a high threshold of complexity *would eliminate* all of its

---

equation that is ubiquitous in mathematical physics! When we later come to approach the works of Connes and Kontsevich, we will see how the richness of the junctures between abstract mathematics and physics is astonishing beyond measure.

236 Lax evokes 'the Kolmogorov-Arnold-Moser Theorem which says that a system near a completely integrable system behaves as if it were completely integrable. Now, what near means is one thing when you prove theorems, another when you do experiments. It's another aspect of *numerical experimentation revealing things.*' (M. Raussen & C. Skau, 'Interview with Peter D. Lax', *Notices of the American Mathematical Society* 53, 2006: 223–9: 224, emphasis ours.) And so, extensive calculations in the quidditas helps to *uncover structure* in the eidos: a position quite close to Grothendieck's commentaries concerning certain concrete cohomological calculations that would provoke the structural emergence of motifs.

237 Ibid., 224.

genuine mathematical content, encoded in diverse and contrasting structural proofs – in diverse and contrasting calculative experiments.

The back-and-forth between calculation and structure, between the physical world and mathematical abstraction, between the quiddital and the eidal, is, for Lax, an indispensable process – one that explains the tremendous *vigor* of mathematics:

> My friend Joe Keller, a most distinguished applied mathematician, was once asked to define applied mathematics and he came up with this: 'Pure mathematics is a branch of applied mathematics.' Which is true if you think a bit about it. Mathematics originally, say after Newton, was designed to solve very concrete problems that arose in physics. Later on, these subjects developed on their own and became branches of pure mathematics, but they all came from an applied background. As Von Neumann pointed out, after a while these pure branches that develop on their own need invigoration by new empirical material, like some scientific questions, experimental facts, and, in particular, some numerical evidence. [...] I do believe that mathematics has a *mysterious*

unity which really connects seemingly distinct parts, which is one of the *glories* of mathematics.[238]

The robust mathematical abstractions and large-scale computations that Lax practices *fold back* into one another. The 'unity' – and the consequent 'glory' – of such transits constitutes one of the *specificities* of advanced mathematics. Moreover, when the transit *between* the 'pure' and the 'applied' – moving openly and without any privileged direction between *both* poles – breaks with the reasonable expectations of the mathematical community, the 'glory' and the 'honor of the human spirit' are intensified. This is the case, as we shall now see, with the 'applications' of the Lax-Phillips semigroup in number theory, and, even more surprisingly, as we shall see later, with Connes's strategy in his (ongoing) effort to find a proof for the Riemann hypothesis by means of instruments drawn from physics, and with his work dealing with the 'appearance' of the Grothendieck-Teichmüller group in cosmology.

In their works concerning the spectrum of an operator over a hyperbolic variety, Lax and Phillips introduce a formal semigroup – a collection of operators $Z(t)$ associated with orthogonal projections of waves over suitable subspaces of the Hilbert space $L_2(R)$ – in order to control

---

238 Ibid., 225 (emphasis ours). Von Neumann, the young Hungarian mathematician's mentor in Los Alamos, is, for Lax, the exemplar to follow in mathematics: a powerful vision and great calculative capacity, always breaking the supposed barriers between pure and applied mathematics.

the scattering associated with the propagations of waves (that is to say, the asymptotic behavior of waves in the remote past or future).[239] Fadeev and Pavlov later observe (1972) that revealing connections with the harmonic analysis of certain automorphic functions emerge when we apply Lax and Phillips's theory to the *non-euclidean* wave equation.[240] Refounding scattering theory on non-euclidean bases, Lax and Phillips go on to characterize the meromorphic properties of Eisenstein series,[241] producing explicit formulas and exhibiting proofs that are both concise and general ('elegant', that is, in Lax's aforementioned sense), and that thereby uncover a *totally unexpected transit* between the differential and the arithmetical, through the properties of the semigroup $Z(t)$.[242]

Nevertheless, in a subsequent revision of the theory,[243] Lax and Phillips explain how the connection between the *natural* non-euclidean geometry modeled on the Poincaré

---

239 P. Lax & R. Phillips, *Scattering Theory* (New York: Academic Press, 1967).

240 Given a wave equation $u_n = c^2 \nabla^2 u$ (with the Laplacian $\nabla^2 = \Sigma \partial^2 / \partial_i^2$), the non-euclidean wave equation is obtained by means of the perturbation $u_n = c^2 \nabla^2 u + u/4$.

241 Given the Poincaré plane z (that is, $z \in \mathbb{C}$ with $Im(z) > 0$) and given $k \geq 2$, the associated Eisenstein series is defined by $\Sigma_{m,n \neq 0} (m + nz)^{-2k}$. This is a holomorphic function that converges absolutely on the Poincaré plane, which turns out to be invariant under the modular group $SL_2(\mathbb{Z})$ and which extends into a meromorphic function over $\mathbb{C}$. The remarkable *Ramanujan identities*, which the ingenious Indian mathematician proposed with respect to the coefficients of the Eisenstein series, are well known, and correspond to sophisticated differential identities between the series.

242 P. Lax & R. Philips, 'Scattering Theory for Automorphic Functions', in *Annals of Mathematics Studies* 87 (Princeton: Princeton University Press, 1976).

243 P. Lax & R. Phillips, 'Scattering Theory for Automorphic Functions', *Bulletin of the American Mathematical Society*, New Series 2, 1980: 261–95.

plane (the group of rational transformations $w \to (aw+b)/(cw+d)$ with $a,b,c,d \in \mathbf{R}$, and $ab-bc=1$ and the various types of *natural* invariants associated with that geometry L_2, Dirichlet, Laplace-Beltrami)[244] turns out to be the deep connection that lets us unfold the *'intrinsic meaning'*[245] hidden in differential equations like the non-euclidean wave equation, a meaning that can be glimpsed precisely in virtue of the semigroup $Z(t)$. In this manner, we observe how a complete *mixture* in Lautman's sense (the Lax-Phillips semigroup) allows us to *naturally mediate* between the (apparently distant) realms of the differential and the arithmetical, thanks to the discovery of a *single natural model* that pertains to both realms. That model is the Poincaré plane, seen as a non-euclidean model, with its differential Riemannian geometry and analytic invariants, on the one hand, and the same plane, seen as a complex model, with its theory of automorphic functions and arithmetical invariants, on the other. In situations of this sort, we confront a sophisticated web of transits between the quiddital and the eidal, with multiple contrasting supports in the web: physical motivations (scattering, waves); concrete models (Poincaré plane, non-euclidean geometry, modular forms), generic structures (geometries, invariants, semigroups).

---

244 Ibid., 262.
245 Ibid.

A sophisticated web of motivations issuing from physics, a very broad spectrum of examples from functorial analysis, geometry and algebra, and a powerful abstract theorematic machinery are combined in the work of **Alain Connes** (France, b. 1947): the concretization of a mathematics profoundly oriented toward the quiddital, but which is also reflected in the (pendular and inevitable) eidal transit – of higher mathematics. A Fields medalist (1982) for his works on the classification of *operator algebras* in Von Neumann algebras, and for his applications of the theory of C^*-algebras[246] to differential geometry, Connes has, from early on (since his doctoral thesis in 1973), worked on the unification of various, apparently distant, abstract conceptual instruments (modular and ergodic operators, projectivity and injectivity properties), and on their manifold uses[247] in functional analysis and the

---

[246] A C^*-algebra is a Banach algebra (an associative algebra with a complete normed topology) with an involution operator ( )* that behaves multiplicatively with respect to the norm. The original examples of C^*-algebras are matrix algebras (linked to Heisenberg's matrix mechanics) and linear operator algebras over Hilbert spaces (linked to quantum mechanics, following Von Neumann). Von Neumann algebras are C^*-algebras of operators that are closed under certain weak topologies. C^*-algebras are *mixed* mathematical objects in Lautman's sense, in which *the linear and the continuous* are interlaced, through a hierarchy of *intermediary* properties that have to do with convexness, order, identities and quotients. For a presentation of Connes's early works, see H. Araki, 'The Work of Alain Connes', in Atiyah, Iagolnitzer, *Fields Medallists' Lectures*, 337–44.

[247] The weaving between the One and the Multiple is *fully bipolar* in Connes. Indeed, he makes use of the abstract instruments of mathematics for applications in physics (the use of C^*-algebras for understanding quantum mechanics, sharpening Von Neumann's program), but also, as we will later see, he makes use of the concrete instruments of physics for 'applications' in mathematics (the use of spectroscopy for understanding the Riemann hypothesis). As regards the evanescent *frontier* between the pure and the applied, recall Keller's paradoxical definition of pure mathematics as a branch of applied mathematics, as evoked by Lax.

underlying mathematical physics. Connes subsequently obtained an index theorem for foliations[248] (1981), and, in the eighties, began to develop his *noncommutative geometry*. In the wake of such great unifying works as those of Von Neumann, Grothendieck and Atiyah, Connes *opens* mathematics onto research programs of tremendous breadth.[249]

The emergence of the *noncommutative paradigm* in Connes rests upon three basic pillars:[250] the real (quiddital) ubiquity of spaces whose coordinate algebras are non-commutative; the technical power of abstract (eidal) instruments that can be extended to noncommutative situations (cyclic cohomology, K-homology, spectral theory,

---

248 A foliation is a differential manifold that is locally decomposed into parallel affine submanifolds (the 'leaves' of the foliation). Foliations appear everywhere in mathematics: given a immersion $f : M \to N$ between varieties with $dim(M) \geq dim(N) = n$, we obtain an n-foliation over M, the leaves of which are the components of $f^{-1}(x)$, $x \in N$; given a Lie group G acting on a manifold M in a locally free fashion (which is to say, such that for all x in M, $\{g \in G : gx = x\}$ is discrete), the orbits of G make up the leaves of a foliation over M; given a nonsingular system of differential equations, the family of solutions to the equation make up a foliation, and the global determination of the solutions determines the behavior of the foliation. An *index theorem for foliations*, such as the one obtained by Connes, therefore interlaces certain general constructions from differential geometry with the underlying techniques of algebraic geometry in the Index Theorem. Advanced mathematics thereby continues to soar over the *incessant transits* between its subdomains.

249 Fittingly, Connes is (in Grothendieck's footsteps) a permanent professor at the IHES (since 1979), and (in Serre's footsteps) a permanent professor at the Collège de France (since 1984).

250 Connes is a magnificent expositor and defender of his ideas. See, for example, A. Connes, *Noncommutative Geometry* (San Diego: Academic Press, 1994); A. Connes & M. Marcolli, *A Walk in the Noncommutative Garden*, preprint, http://arxiv.org/abs/math/0601054. Connes's website (www.alainconnes.org) has a complete bibliography, available as an archive of pdf files. The descriptions appearing in our text come from Connes's writings, with certain emphases added by us.

the 'thermodynamics' of operators); the harmonic richness of certain very general *motivations*:

$$\frac{\text{Euclidean Geometry}}{\text{Non-euclidean Geometry}} \equiv \frac{\text{Commutative}}{\text{Noncommutative}} \equiv \frac{\text{Terrestrial Physics}}{\text{Cosmic Physics}}$$

In fact, we see noncommutativity appear *in a natural manner* in essential areas of physics (phase spaces of quantum mechanics, cosmological models of space-time), geometry (duals of discrete non-abelian groups, non-abelian tori, foliation spaces), and algebra (adele spaces, modular algebras, Q-lattices). Here we discover a certain ubiquity of the noncommutative in *actual nature*,[251] which goes hand in hand with an *extension of the notion of* space from the point of view of its *conceptual nature*: the passage from infinitesimal manifolds (Riemann) to C^*-algebras of compact operators (Hilbert, Von Neumann); the passage from dual K-homology (Atiyah, Brown, Douglas, Filmore) to noncommutative C^*-algebras (Connes); the passage from the Index Theorem (Atiyah, Singer) to the handling of noncommutative convolutions in groupoids (Connes); the passage from the groups and algebras of modern differential geometry (Lie) to quantum groups and Hopf algebras;[252] the passage from set-theoretic

---

251 It would not be out of place here to recall Lautman's remarkable study of *symmetry and dissymmetry* in mathematics and physics, as a sort of prelude to subsequent noncommutative studies (see p. 59, above).

252 Hopf algebras are structures that show up in the proofs of representation theorems for algebraic groups (combinations of groups and algebraic varieties, such as linear groups, finite groups, elliptic groups, etc.). Vladimir Drinfeld (1990 Fields medalist) introduced quantic groups (1986) as nonrigid deformations of Hopf algebras, and

punctuality to the actions of noncommutative monoids in Grothendieck topoi, etc.

Constrained as we are to make a selection from the many and various results and subprograms of investigation brought forward by Connes in his approach to the quiddital, let us emphasize two of them here: the emergence of a 'cosmic' Galois group that is close to the 'absolute' Galois group in number theory, a form of transit between a well-known eidal configuration (absolute group) and an unexplored quiddital one (cosmic group); the utilization of spectroscopic techniques in an effort to demonstrate the Riemann hypothesis, an *inverse* form of transit between the quiddital and the eidal. In a famous article, Pierre Cartier, one of Grothendieck's main disciples, had conjectured that 'there are many reasons for believing in a "cosmic Galois group" acting on the

---

showed that they appear naturally in the Yang-Baxter equation, a pivotal equation for domains of statistical mechanics. In turn, string theory in contemporary physics – the Pascalian utopia of a harmony between the infinitely small (quantum mechanics) and the infinitely large (general relativity) – stands in need of a sophisticated mathematical theory of knots, which can be adequately handled only by means of quantic groups and *n*-categories (categories in which one *climbs the scale of transits* – beyond morphisms between morphisms [functors], and morphisms between functors [natural transformations], one studies morphisms between natural transformations, and then morphisms between those lower-level morphisms, and so on). Drinfeld's early works (written at the age of twenty!) resolved the Langlands conjecture for the $GL(2:k)$ case, with k being a global field of finite characteristic. As we will see, Drinfeld also proposed a combinatorial description of the Grothendieck-Teichmuller group, with surprising applications in physics. Beginning with Drinfeld, and perhaps culminating in Kontsevich, the Russian school has generated an extraordinary profusion of theoretical results in physics, making use of both the categorical abstractions in Grothendieck's work, and the functorial conjectures of the Langlands program. A *supreme and categorical interlacing* of arithmetic-algebra-geometry would thus seem to be hidden in the continuous mysteries of physics. The philosophical consequences of such a situation are of enormous relevance, but are nevertheless *invisible* in the usual treatments of the philosophy of mathematics.

fundamental constants of physical theories. This group should be closely related to the Grothendieck-Teichmüller group.'[253]

One of those reasons consists in a result of Connes and Kreimer's (1999), in which it is demonstrated that the Lie algebra of the Grothendieck-Teichmüller group[254] acts naturally on the algebra corresponding to Feynman diagrams. Shortly thereafter, Connes and Marcolli (2004) succeeded in demonstrating that the cosmic Galois group can be described as the universal symmetry group U of renormalizable physical theories, and that, in effect, its Lie algebra can be extended to the Lie algebra of the Grothendieck-Teichmüller group.[255]

---

253 P. Cartier, 'A Mad Day's Work: From Grothendieck to Connes and Kontsevich. The Evolution of Concepts of Space and Symmetry', *Bulletin of the American Mathematical Society* (New Series) 38, 2001: 407. Cartier's article was originally written in French in 1998; in 2000, he added a postscript, from which we have taken our quotation. The *absolute* Galois group is the Galois group of the (infinite) algebraic extension $Gal(\bar{\mathbf{Q}}:\mathbf{Q})$ where $\bar{\mathbf{Q}}$ is the algebraic closure of the rationals. The Grothendieck-Teichmüller group (GT) offers a *combinatorial* description of the absolute Galois group. It remains an open conjecture whether or not the algebraic and combinatorial descriptions are equivalent ($GT \stackrel{?}{=} Gal(\bar{\mathbf{Q}}:\mathbf{Q})$). The Grothendieck-Teichmüller group appears in a natural manner in Grothendieck's dessins d'enfants (1983): finitary objects aimed at characterizing the behavior of number fields through certain associated Riemann surfaces. Yet to be fully understood, dessins d'enfants – forms of the combinatorial understanding of algebra by way of analysis – constitute a typical *Grothendieckian transit*.

254 The Lie algebra of GT can be described as a free algebra over the Euler numbers $\zeta(3)$, $\zeta(5), \zeta(7), \dots$ (where $\zeta(k)=\Sigma_{n\geq 1} n^{-k}$). The Euler numbers appear in many corners of number theory, but are still almost entirely unknown objects: only the irrationality of $\zeta(3)$ has been proven (Apéry 1979, a tour de force that remained isolated for many years), and, recently, the irrationality of infinitely many $\zeta(k)$ when k is odd (Rivoal 2000). See Cartier, 'A mad day's work…', 405–6.

255 The Lie algebra of U is the free algebra over $\zeta(1), \zeta(2), \zeta(3), \dots$. The proximities between the absolute Galois group $Gal(\bar{\mathbf{Q}}:\mathbf{Q})$ and the cosmic Galois group U, through the mediation of GT, thus allow us to pinpoint a *totally unexpected action of an arithmetical group on the universal constants of physics* (the Planck constant, the speed of light, the gravitational constant, etc.).

'Cartier's dream', as his conjecture was known for some years, had therefore 'come true', thanks to the results of Connes and his team, and it thus represents a sort of *intensified, infinitely refined Pythagoreanism*, harking back to the first, rough and original hypotheses regarding the existence of harmonic correspondences between *mathematika* (the study of quantity) and *kosmos* (order).[256] In Connes's work, this arithmetico-geometrico-physical refinement extends to deeper analogies between physical divergences in field theory and arithmetical mixtures in Tate motifs,[257] thus approaching what Connes calls the very 'heart' of mathematics: 'modular forms, L-functions, arithmetic, prime numbers, all sorts of things linked to that'.[258] *From the dream to the heart*, we therefore achieve a progressive revelation in the order of discovery, which we have already carefully observed in Grothendieck's work, and which Connes takes up for his art as well: 'There are several phases leading to the "finding" of new math, and while the "checking" phase is scary and involves just

---

256 It would not be out of place here to make a fresh return to Plato's *Timaeus*. *Independently* of the calculations contemplated there, which have obviously been outstripped, Plato's *underlying relational strategy* does not turn out to be too far from the relational search for correspondences between arithmetico-geometrical forms and cosmological structures that is now being contemplated by Cartier, Connes and Kontsevich. We will return to these questions in this work's third part.

257 Tate's mixed motifs (1965) show up in the representation of the homology classes of a variety by means of linear combinations of subvarieties (algebraic cycles) and in the connections between that representation and *l*-adic cohomology. Tate motifs serve as a concrete guide to Grothendieck's general conjectures concerning motifs (standard conjectures).

258 C. Goldstein & G. Skandalis, 'An Interview with Alain Connes', *EMS Newsletter* 63, 2007: 25–30: 27.

rationality and concentration, the "creative" phase is of a totally different nature'. The emergence of *simple ideas* after very *lengthy experimentations* and the transit through 'mental objects which represent *intermediate steps* and results at an idealized level' underpin the specificity of doing mathematics.[259]

Connes's inventiveness reaches a still-higher pitch in his program to demonstrate the Riemann hypothesis by means of strategies and techniques drawn, essentially, from physics.[260] The passage from the quiddital to the eidal that surrounds the Riemann hypothesis is supremely original.[261] On the one hand, Connes points out that a quantum-theoretical study of the *absorption spectrum of*

---

259 Alain Connes, 'Advice to the Beginner', http://www.alainconnes.org/docs/Companion.ps, 2.

260 Connes shows himself to be a devoted lover of formulas and calculations. See Goldstein & Skandalis, 'An Interview with Alain Connes' 28.

261 The Riemann hypothesis encodes certain arithmetical properties in terms of analytic properties. Riemann's zeta function is a function of a complex variable that is initially defined by means of an absolutely convergent series $\zeta(s) = \sum_{n \geq 1} n^{-s}$ for the case where $Re(s) > 1$ (over natural numbers greater than 1, this therefore coincides with the Euler numbers) that subsequently, through an analytic extension, gives rise to a meromorphic function over $\mathbf{C}$, with a simple pole in $s = 1$ (residual 1). A functional equation obtained by Riemann for the zeta function shows that it possesses 'trivial' zeros (roots) in the negative even integers. Riemann conjectured (1859) that *all the other zeros* of the zeta function lie in the complex line $Re(z) = 1/2$ (the so-called Riemann hypothesis). Various mediations serve to link Riemann's zeta function to arithmetic: the Euler formula $\sum_{n \geq 1} n^{-s} = \prod p$ prime $1/(1-p^{-s})$; other 'mixed' functions of a complex variable determined by the zeta function; intermediary functional equations between these; and the subtle asymptotic behavior of the functions. Riemann's strategy inaugurates a profound understanding of the *discrete* through underlying *continuous* instruments, which would go on to be furthered by the German school of abstract algebra (Artin, Hecke), and which would give rise to the Weil conjectures and to Grothendieck's grand cohomological machinery. The consequences of the Riemann hypothesis in number theory are quite extensive, and the hypothesis may perhaps be considered, today, as the greatest open problem in mathematics. For a description of the situation, see E. Bombieri, 'The Riemann Hypothesis', in J. Carlson et al., *The Millennium Prize Problems* (Providence: The Clay Mathematics Institute, 2006), 107–24.

light, using the instruments of noncommutative geometry, allows us to recalculate, with all the desired precision, all of the constants that appear in the limited developments of Riemann's zeta function.[262] A fundamental critical turn in that approach consists in calibrating the appearance of a negative sign (which Connes qualifies as 'cohomological') in the approximations of the zeros of the zeta function through absorptions (and not emissions) of a spectrum. On the other hand, Connes offers a sweeping construction of *analogies* in an attempt to *transfer* to the case of finite extensions of **Q** ('number fields') Weil's 1942 demonstration of the generalized Riemann hypothesis for global fields of characteristic $p > 0$.[263] Connes's strategy (announced in 2005 with Consani and Marcolli, and with an eye to the near future) here consists in progressively eliminating the *obstructions* in the transit by means of an elucidation of concepts, definitions and techniques in noncommutative geometry that correspond to Weil's successful undertakings in algebraic geometry.[264] In this

---

262 A. Connes, 'Trace Formula in Noncommutative Geometry and the Zeros of the Riemann Zeta Function', *Selecta Mathematica* (New Series) 5, 1999: 29–106. The idea of using the spectrum and the trace of an operator in a suitable Hilbert space to capture the zeros of the zeta function comes from Hilbert and Pólya, as Connes himself indicates. His originality consists in *combining* the *natural* instruments of noncommutative geometry connected to Hilbert spaces with the *universal* physical situations underlying those instruments.

263 For details see Connes & Marcolli, *A Walk in the Noncommutative Garden*, 84-99.

264 Connes's program clearly shows how certain webs of invention and discovery proceed in higher mathematics. The *analogies* – or harmonious *conjectures* – correspond to precise (but not theorematic) translations between algebraic geometry and noncommutative geometry, with transferences and technical redefinitions of concepts in either context. The refined structural organization of each realm allows for an intuition of synthetic

way, we may gain access to characteristic zero as the 'limit' (in noncommutative geometry) of those that are well behaved in characteristic p. In each of these processes, *a back-and-forth between the quiddital and the eidal, without any privileged direction being fixed in advance*, allows for the emergence of mathematical results of great depth – a depth that is not only conceptual and technical but also philosophical.

Maxim Kontsevich (Russia, b. 1964) is another remarkable contemporary mathematical author who has been able to unite high speculative abstraction and the concrete richness of physical phenomena. Kontsevich was awarded the Fields Medal in 1998. In his acceptance speech, he said,[265]

> For myself, as a mathematician, it is very interest-
> ing to decipher the rules of the game of theoretical
> physics, where one doesn't see structures so much
> as the symmetry, locality and linearity of observ-
> able quantities. It is very surprising that those weak

correspondences, which are then analytically delimited and contrasted with the many examples available, thereby producing a sort of *dictionary* between algebraic geometry and noncommutative geometry. A series of 'analogies' can be found in Connes & Marcolli, *A Walk in the Noncommutative Garden*, 9. As the authors indicate, the fluctuations implicit in the analogies are what *drives the subsequent development* of the mathematics. To eliminate this *indispensable, initial vagueness* from mathematics – as a century of analytic philosophy presumed to do – therefore impedes understanding of the complex creative forms of the discipline.

265 For a technical description of Kontsevich's work prior to the Fields Medal, see C. H. Taubes, 'The Work of Maxim Kontsevich', in Atiyah, Iagolnitzer, *Fields Medalists' Lectures*, 703-10.

conditions ultimately lead to such rich and complicated structures.[266]

This is the case with *Feynman diagrams* in theoretical physics,[267] whose formal *use* in mathematics was introduced by Kontsevich in order to resolve some formidable problems: Witten's conjecture regarding moduli spaces of algebraic curves; the quantization of Poisson varieties; and the construction of invariants of knots.

In the arithmetic of moduli spaces of algebraic curves, certain cohomological invariants appear ('intersection numbers'), which may in turn be represented as complex combinatorial coefficients of a formal series $F(t_0\bar{t})$. Formally manipulating two quantum field theories, Witten conjectures that the formal series $U=\partial^2 F/\partial t_0^2$ should satisfy the KdV equation,[268] which would give rise to numerous interrelations between the intersection numbers in the arithmetic of algebraic curves. Making use of his great talent for combinatorial calculations, Kontsevich

---

266 M. Kontsevich, reception speech at the Académie des Sciences (2003), http://www.academie-sciences.fr/membres/K/Kontsevich_Maxim_discours.htm, 2.

267 Feynman diagrams are graphs by which one may represent the perturbations of particles in quantum field theory. Certain combinatorial (dis)symmetries and (dis)equilibria in the diagrams no only suffice to simplify the calculations but *predict* new physical situations, which subsequent mathematical calculations confirm. The heuristic use of the diagrams in theoretical physics has been very successful, opening an important gateway onto *visuality* in theoretical knowledge. A *mathematical formalization* of the diagrams appears in A. Joyal & R. Street, 'The Geometry of Calculus I', *Advances in Mathematics* 88, 1991: 55-112 (using the instruments of category theory). On the other hand, the *mathematical use* of the diagrams in order to resolve deep mathematical problems is due to Kontsevich.

268 See p. 215, n. 235.

proved Witten's conjecture and succeeded in exhibiting those interrelations in an *explicit* fashion, starting with constructive models for moduli spaces, based on Riemann surfaces of diagrams with metrics.[269] This is an extraordinary example of the inventive richness of contemporary mathematics, whereby arithmetic and physics are woven together in ways that are *not predetermined* in advance: the conjecture is arithmetico-differential, motivated by a comparison with physics, and the proof interlaces combinatorial, arithmetical and continuous fragments on the basis of certain physical images (diagrams, graphs, surfaces, metrics). The *bipolar transit* between physics and mathematics is thus truly the generator of a new knowledge. What matters here is not a *supposedly originary base* (the physical world, the world of mediations or the world of ideas) that would firmly support the edifice of knowledge, but a *tight correlational warp* that supports the transit of knowledge (see chapters 8 and 9).

Phenomena of quantization – the deformation of observable quantities by means of new parameters and the asymptotic study of the deformations as the parameters tend toward zero – show up on many levels in physics, and, in particular, in the study of the infinitely large and the infinitely small. On the one hand, in general relativity, it has been observed how the Poincaré group

---

269 M. Kontsevich, 'Intersection Theory on the Moduli Space of Curves and the Matrix Airy Function', *Communications in Mathematical Physics*, 147, 1992: 1-23.

(isometries of Minkowski space-time) tends toward the Galileo group (isometries of euclidean space-time) when the parameter bound to the *speed of light* tends toward zero. On the other hand, in quantum mechanics, it has been observed how the 'natural structures' of quantum mechanics tend toward the 'natural structures' of classical mechanics when the parameter tied to the *Planck constant* tends toward zero. The structures of classical mechanics are well known, and correspond to the *Poisson manifolds*,[270] in which one can naturally formalize the Hamiltonian as an operator for the measurement of the energy (well-determined energy and momentum) of classical physical systems. Although, from a mathematical point of view, the quantizations of an algebra have, since the 1950s, been understood as quotients of formal series over that algebra (Kodaira), the *quantizations of Poisson manifolds* (which later showed up in quantum mechanics) had not been rigorously studied before Kontsevich. Here, Kontsevich discovers that such a quantization is linked to a *new kind of string theory,* where the use of Feynman diagrams is significant; and he exhibits, again in an *explicit* manner, that the deformation is linked to certain perturbations of quantum fields and to extremely subtle calculations

---

270 A Poisson algebra is an associative algebra with a Lie bracket that acts as a *derivation* of the algebraic operation (the law $[x, yz] = [x, y]z + [x, z]y$, is to be read as an analogue of 'Leibniz's Law': $\partial_x(yz) = (\partial_x y)z + (\partial_x z)y$). A Poisson manifold is a differential manifold with the structure of a Poisson algebra. The paradigm of a Poisson manifold is the algebra of smooth functions over a *symplectic manifold* (a generalization of a manifold with a Hamiltonian).

of certain terms of asymptotic expansions.[271] Even more astonishingly,[272] *what emerges in Kontsevich's calculations* is an action of the Grothendieck-Teichmüller group on the space of possible universal formulas of physics, a group that can also be seen as the symmetry group of the possible quantizations of the original Poisson manifold. And so we find, in a totally unexpected way, a corroboration of Connes's simultaneous discovery regarding the action of the absolute Galois group on the universal constants of physics.

A third unexpected place where Feynman diagrams show up, to help resolve highly sophisticated mathematical problematics, is in the construction of universal invariants in mathematical knot theory.[273] In his work, Kontsevich introduces an entire series of novel constructions over which the invariants are layered: complex differentials of

---

[271] M. Kontsevich, 'Deformation Quantization of Poisson Manifolds I', *IHES preprints* M/97/72 (1997).

[272] Kontsevich describes the 'surprise' caused by the emergence of new strings in his quantization calculations (Kontsevich, speech, *Académie des Sciences*, 1). The surprise caused by the action of the Grothendieck-Teichmüller group was in no doubt greater still.

[273] A mathematical knot corresponds to the intuitive image of a string tied into a knot, with its extremities identified (formally, a knot is therefore an immersion of the circle S^1 in R^3). A general and complete classification of knots is still an open problem. Poincaré, Reidemeister and Alexander, in the first half of the twentieth century, proposed some initial instruments for such a classification. But it was above all in the last decades of the twentieth century, with the work of Jones and Witten (1990 Fields medalists), that knot theory made its theoretical breakthrough. Vassiliev proposed a series of topological invariants connected to the Jones polynomial, which Kontsevich had reconstructed by means of abstract *integrals* over suitable algebraic structures, with strong *universal properties*. See M. Kontsevich, 'Feynman Diagrams and Low-Dimensional Topology', *First European Congress of Mathematics (Paris 1992)* (Boston: Birkhäuser, 1994), 97–121, and M. Kontsevich, 'Vassiliev's Knot Invariants', *Advances in Soviet Mathematics*, 16/2, 1993: 137–50.

graphs; cohomology groups of those complex differentials, differential forms over those groups; integrability of those forms through a generalized Stokes argument; etc. And so we find ourselves in the company of a mathematician extraordinarily skilled in combinatorial manipulations, and endowed with tremendous plasticity in the most diverse forms of *exact transit*: the technical and calculative passage between nearby subbranches of mathematics; the analogical and structural passage between more distant realms of mathematics; and the visual and conceptual passage between mathematics and physics.

We encounter another example of Kontsevich's transgressive power in his ideas for homologically formalizing the phenomena of *mirror symmetry* in theoretical physics.[274] Witten described a topological unfolding in supersymmetry phenomena, which corresponds to a sort of specular reflection between strings (A-branes and B-branes, sophisticated models that incorporate Riemann surfaces and holomorphic manifolds). Kontsevich conjectured that the mirror symmetry between two manifolds X, Y should correspond to an *equivalence of two triangulated categories*,[275] one coming from the *algebraic*

274 M. Kontsevich, 'Gromov-Witten Classes, Quantum Cohomology and Enumerative Geometry', *Communications in Mathematical Physics*, 164, 1994: 525–62; M. Kontsevich & Y. Soibelman, 'Homological Mirror Symmetry and Torus Fibrations', in K. Fukaya et al., *Symplectic Geometry and Mirror Symmetry* (Singapore: World Scientific, 2001), 203–63.

275 Triangulated categories supply axioms (the naturalness of which is still under discussion) that can be put to use in trying to universally capture the properties

geometry of X, and the other from the *symplectic geometry* of Y. In this manner, complex symmetry phenomena in the physics of the infinitely small correspond to transfers of structure between the discrete (algebraic varieties) and the continuous (symplectic manifolds), thus giving rise to another new, totally surprising and unexpected connection between physics and mathematics. Kontsevich's hypothesis was later *mathematically* demonstrated in many cases – for elliptic curves (Kontsevich, Polischuk, Zaslow), for tori (Kontsevich, Soibelman), for quartics (Seidel) – and *physically* confirmed with the discovery of new strings (D-branes) anticipated by the theory.

Kontsevich's other works explore very deep connections between motifs, 'operads',[276] the cohomology of Lie algebras and the topology of varieties, seeking to provide the foundations for a ubiquitous *quantum cohomology*, which would reveal the presence of certain universal

---

of the *derived category* of an abelian category. Given an abelian category A (which generalizes the properties of the category of abelian groups), $Com(A)$ is the category of its simplicial complexes with chain morphisms, and $Der(A)$ is the derived category whose objects are homotopy classes of the objects of $Com(A)$, and whose morphisms are localizations (modulo quasi-isomorphism) of the morphisms of $Com(A)$. $Com(A)$ and $Der(A)$ are triangulated categories. The notion comes from Grothendieck and Verdier (at the beginning of the sixties, and with Verdier's 1967 thesis – not published until 1996!), where it is used to express certain properties of duality in a general fashion.

276 *Operads* can be understood in terms of the analogy: algebras/operads ≡ representations/ groups. Operads are collections of operations that compose with one another well and that realize a sort of minimal (*disincarnate*) compositional combinatorics, underlying the higher algebras in which the operads are *incarnated* in a concrete fashion. Operads can thus be seen as one more example of generic – or *archetypical* – constructs, like many other mathematical objects that we have been parading through these pages (universal objects in categories, cohomologies, motifs, possible cofinalities, etc.). In the next chapter we will supply additional mathematical facts concerning this emergence of 'archetypes', and in chapters 8 and 9 we will study their ontic and epistemic status.

algebraic 'archetypes' behind the many continuous functions of physics.[277] At stake here is a situation one may find at the very 'center' of mathematics, and which would answer, in a novel manner, to what Thom called the 'founding aporia' of mathematics. Indeed, Kontsevich has explicitly pointed to a possible enlarging of the *heart* of mathematics (recalling Connes), in which the radical importance of the current connection between physics and mathematics is emphasized:

> The impact that the new discoveries in physics have had on mathematics is enormous. One could say that, before, in mathematics, there existed a principal center of mysteries, the group of all the conjectures interlacing number theory, the motifs of algebraic varieties, the L-functions (generalizations of Riemann's zeta function) and automorphic forms, which is to say the harmonic analysis of a locally homogenous space. Now, however, the theory of quantum fields and string theory constitute a second center of mysteries, and offer a new depth and new perspectives to mathematics.[278]

---

277 M. Kontsevich, 'Operads and Motives in Deformation Quantization', *Letters in Mathematical Physics*, 48, 1999: 35–72; M. Kontsevich, 'Deformation Quantization of Algebraic Varieties', *Letters in Mathematical Physics*, 56, 2001: 271–94.

278 Kontsevich, speech at the Académie des sciences, 1.

In this manner, the quidditas imposes its massive imprint on the eidal signs aimed at helping us understand the world. Amid all of these tensions, we see how the 'world' consists in a series of data/structures (Peircean firstness), registers/models (Peircean secondness) and transits/functors (Peircean thirdness), whose *progressive interlacing into a web* not only allows us to better understand the world, but *constitutes* it in its very emergence. In the next chapter ,we will see how contemporary mathematics is finding new, stable supports in that web (invariants, 'archetypes'), *thus solidifying the relational and synthetic gluing of both phenomena and concepts*, without any need of an analytic foundation to ensure the security of the transit.

CHAPTER 7

ARCHEAL MATHEMATICS: FREYD, SIMPSON, ZILBER, GROMOV

A metaphor for understanding the complex transits that take place in contemporary mathematics can be found in the image of an *articulated pendulum*. Unlike a simple pendulum, which, as it sweeps its course, determines a *fixed* frontier, equidistant between its extremes, an articulated pendulum – built by linking together two pendula oscillating in *opposite* directions – defines an altogether extraordinary *dynamic* curvature, unimaginable if one were just to consider the two pendula separately. In fact, in a chronophotograph of an articulated pendulum by Marey (1894 – see figure 10, overleaf), we can see how an entire *extensive spectrum of the intermediate* emerges in the reticulated undulation to the left, thereby opening itself to the curvatures characteristic of the living and the organic. The contrast between an articulated and a simple pendulum serves as a simple, metaphorical contrast between advanced and elementary mathematics. Indeed, on the one hand, advanced mathematics – and especially contemporary mathematics, as we have shown in the second part of this work – harbors an entire series of dialectical concretizations arising from a sophisticated *articulation* between webs and scales of concepts and models, on multiple eidal and quiddital levels.

On the other hand, the low levels of complexity in the techniques of elementary mathematics *inherently* simplify the underlying conceptual movement; they therefore have no need of truly subtle (infinitely discerning) articulations or hierarchies in order to erect their edifice. We thus have a metaphorical counterpoint between the articulated and the simple, which is further supported by a counterpoint between a 'relative mathematics', a mathematics in motion, in the style of Grothendieck (contemporary mathematics, articulated pendulum), and an 'absolute mathematics', a mathematics at rest, in the style of Russell (elementary mathematics, analytic foundation, simple pendulum).

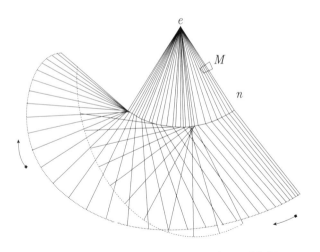

Figure 10. *Articulated pendulum, after a chronophotograph by Marey.*

The (dynamic, organic, living) curvature we see on the left side of figure 10 may seem to *transcend* the composite and the projective, while the situation can actually be reconstructed on the basis of *prototypical* underlying tensions (the contrary impulses between the greater and the lesser pendula) by anyone who knows how the articulated pendulum works. Likewise, behind the processes of ascent and descent that we have been describing in the preceding chapters, behind the pendular oscillations between fragments of ideality and reality, behind what we have called the *bipolar* dialectic between the eidal and the quiddital – which is to say, behind the *incessant transit in both directions* between concepts and data, between languages and structures, between mathematics and physics, between imagination and reason – contemporary mathematics has gone on to produce *deep archetypes*, by which the transit can be *stabilized*, the opposing polarities *mediated*, and the pendular movements *balanced*.

With the neologism *archeal* (from *archê* [principle or origin, see figure 11, overleaf]) we will designate, in this chapter, the search for (and discovery of) remarkable *invariants* in contemporary mathematics, by which the transits can be soundly controlled, *without any need of being anchored in an absolute ground*. Those invariants will serve as *relative* origins (*arkhô*), commanding (*arkhên*) movement on a particular level (that is, in specific concrete categories). We will therefore take up a revolutionary conception which has surfaced in contemporary mathematics *in a theorematic manner:* the register of *universals capable of unmooring themselves from any 'primordial' absolute, relative universals* regulating the *flow* of knowledge. In this chapter we will discover certain technical constructions in that register of 'decantations of the universal', and in chapter 9 we will inquire as to how we may contain the apparent contradiction in terms 'relative universal', which will give rise to a new *founding synthetic aporia* of mathematics (one that is nonanalytic – that is, non*foundational*).

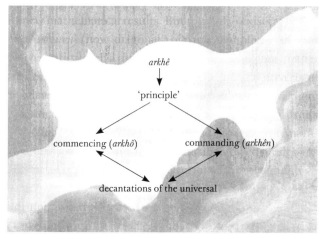

Figure 11. *The Archeal Realm.*

The works of **Peter Freyd** (USA, b. 1935) in category theory forcefully bring to light the emergence of archetypes in the structuration of mathematical thought. As we have already seen with Grothendieck, the dialectic of the One and the Many finds one of its happiest expressions in categorical thought, where *one* object defined by means of universal properties in abstract categories will in turn appear or turn out to be *many*, over the plurality of concrete categories where it 'incarnates'. The One and the Universal enter into perfect counterpoint and dialogue with the many and the contextual. Going one step further, Freyd's *allegories*[279] are abstract categories of *relations*, axiomatized through a generic relational combinatorics, beyond the very restrictions of functionality. Here, full attention to the categorical diagrams of relational composition serves to reveal the precise mechanisms of *one/many* adaptation that mediate between logical theories and their various representations – mechanisms that a functional, set-theoretic reading would fail to detect.

Indeed, Freyd's categorical and relational machinery furnishes axiomatizations for hitherto unnoticed *intermediate* categories – categories of lesser representational power than Lawvere's topoi – and shows that they constitute classes of natural models for intermediate logics between certain 'minimal' logics and intuitionistic logic. With the discovery of a remarkable, *ubiquitous* process in categorical logic, which we will turn to in a moment,

279 P. Freyd & A. Scedrov, *Categories, Allegories* (Amsterdam: North-Holland, 1990).

243

Freyd shows how, starting from pure type theories with certain structural properties (regularity, coherence, first-order, higher-order), one can *uniformly* construct, by means of a controlled *architectonic hierarchy*, *free* categories that reflect the given structural properties in an origin (regular categories,[280] prelogoi,[281] logoi,[282] and topoi[283]). In reaching the free categories, we obtain the most 'disincarnate' categories possible, categories that can be projected into *any other* category with similar properties: Freyd thereby succeeds in constructing something like *initial archetypes* of mathematical theorization. His discovery is doubly significant, because it not only describes the invariants of logico-relational transit, but seizes them in a universal manner, *beyond* the particular fluctuations of each logical fragment. As often happens with great turns in mathematics, Freyd's results will only be fully understood in the future, but it is already easy to predict their extraordinary importance.

---

280 Regular categories are categories with the exactness properties (cartesianity, existence of images, preservation of coverings under pullbacks) that are necessary and sufficient for achieving an adequate *composition* of relations.

281 Prelogoi are regular categories for which the subobject functor takes values in the category of lattices (and not only in sets). A preorder P, understood as a category, will turn out to be a prelogos if and only if P is a distributive lattice with a maximal element.

282 Logoi are prelogoi for which the subobject functor (with respect to its values in the category of lattices) possesses a right adjoint. A preorder P, considered as a category, will turn out to be a logos if and only if P is a Heyting algebra.

283 We have already witnessed the appearance of topoi in Grothendieck's work and its subsequent elementary axiomatization due to Lawvere. The category P associated with a preorder will turn out to be a topos if and only if P reduces to a point.

Freyd's procedure begins by taking a given logical theory, and then goes on to capture, *by way of an intermediate free category of relations*, the free terminal category of morphisms that faithfully represents the properties of the initial theory. The structure of the process[284] is T(theory) $\rightarrow A_T$(allegory) $\rightarrow MapSplitCor(A_T)$(category), which yields a free result when one starts with a pure type theory, and which shows, in each of its stages – relationality, subsumption in identity (Cor), partial invertibility ($Split$), functionality (Map)[285] – how a determinate mathematical conglomerate goes on to be 'filtered'. Out of this 'filtration' come two observations of great interest, both mathematically *and philosophically*: 1. the *analytic* process of decomposing the transit is linked to the exhibition of a universal *synthetic* environment that emerges in the process (the allegory A_T), once again stressing the existence of an indispensable[286] analytic-synthetic dialectic

---

284 Ibid., 277.

285 *Correflexivity* generalizes (in the axiomatic environment of allegories) the property of a relation being contained in a diagonal (basic example: partial equivalence relations, 'PERS', which are increasingly used in computability theory). *Cor* functorially captures correflexivity. *Partial invertibility* generalizes the invertibility property of morphisms to the right (basic examples: regular elements in a semigroup, sections of a sheaf). *Split* functorially captures partial invertibility. *Functionality* generalizes (as always, in the allegorical environment) the usual set-theoretic restriction of functionality on relations. *Map* functorially captures functionality.

286 That dialectical indispensability turns out to be *necessary* in the context of Freyd's representation theorems. The fact that delicate philosophical problems depend upon partial *theorematic reflections* in contemporary mathematics is one of the great strengths of those advanced mathematics, as compared to elementary mathematics, in which, *for lack of complexity*, such reflections do not appear. The low threshold of complexity in elementary mathematics therefore turns out to be a true *obstruction* from a philosophical point of view. We will come back to these questions in Part Three.

in mathematics; 2. *beyond* the terminal bipolar objects in a one-to-one correspondence with one another (theories and regular categories, coherent theories and prelogoi, first-order theories and logoi, higher-order theories and topoi), the richness of Freyd's procedure consists in its *progressive adaptation of synthetic, intermediary machinery* (its explication of the actions of the *Cor*, *Split*, *Map* functors).

A *natural search for archetypes* is essential to categorical thought. The various levels of categorical information (morphisms, functors, natural transformations, n-morphisms, etc.), on every level, allow for sub-definitions of *free objects* – universal projective objects (archetypes: 'commanding origins', etymologically speaking) – with far more general universal constructions emerging over each level's projections. This, according to Freyd, is what happens with the process $T \to A_T \to MapSplitCor(A_T)$, as well as with Yoneda's famous lemma[287], which allows us to embed *any* small category into a category of presheaves:

---

[287] Freyd recalls that the lemma does not actually appear in Yoneda's original article ('Note on Products in *Ext*', *Proceedings of the American Mathematical Society*, 9, 1958: 873–75), but in 'a talk that Mac Lane gave on Yoneda's treatment of higher *Ext* functors' (see http://www.tac.mta.ca/tac/reprints/articles/3/foreword.pdf, p. 5). The lemma's immense philosophical content (to which we will return in the book's third part) corroborates the Lautmanian transit between 'notions and ideas' and 'effective mathematics'. It is interesting to point out that the lemma – so close to the structural grounds of Lautman's thought – in fact arose through a vivid discussion between Yoneda and Mac Lane in the Gare du Nord in Paris (see http://www.mta.ca/~cat-dist/catlist/1999/yoneda), so close to the French philosopher's own physical neighborhood.

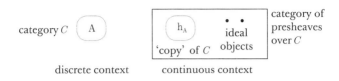

Figure 12. *Yoneda's lemma.*

The representable functors h_A capture the 'coronas' of morphisms surrounding A ($h_A = Mor_C(A, --)$, A being an object in C), but, in the Yoneda embedding ($A \rightarrow h_A$), many other nonrepresentable functors ('ideal' presheaves) show up to *complete* the landscape. In fact, the category of presheaves over C can be seen as a context of *continuity* into which the initial discrete category is inserted (and by which it is completed), since, on the one hand, the category of presheaves possesses all the (categorical) limits, and, on the other, representables preserve limits. The universality of the Yoneda embedding (which obtains for *every* small category) brings two facts of enormous significance to light: *mathematically*, it demonstrates how forms of an *archetypical continuum* lie beneath many discrete situations, forms composed of ideal objects that allow us to reinterpret the initial discrete context (something Hilbert had already predicted in his brilliant text, *On the Infinite*,[288] and which we have found to be carefully articulated,

288 D. Hilbert, 'On the Infinite' (1925), in J. van Heijenhoort, *From Frege to Gödel: A Source Book in Mathematical Logic 1879–1931* (Cambridge: Harvard University Press, 1967), 367–92.

for example, in Langlands's program); *philosophically*, it shows how, in the philosophy of advanced mathematics, the imbrication of fragments of both realism and idealism is not only possible but *necessary* (the *archetypical* completion of 'real' representable presheaves by means of non-representable, 'ideal' ones, and the need for the latter in order to fully understand the former).

In revealing the archetypical nuclei of mathematical thought, the reverse mathematics program of **Stephen Simpson** (USA, b. 1945) takes on a special significance. Initiated in collaboration with Harvey Friedman,[289] the program tries (successfully) to locate *minimal and natural subsystems* of second-order arithmetic that are *equivalent* to the ordinary theorems of mathematical practice derived from those axioms. The equivalence is complete since, in the eyes of adequate underlying theories, the theorems in turn entail the conglomerate of axioms by which they are proved, and so an entire dialectic of deduction and *retroduction* (whence the name: 'reverse mathematics') exists in *demonstrative transit*. There are no fixed foundations in this back-and-forth (it is not a question of an absolute foundation from which everything else would, presumably, follow) but *multiple relative proofs of equideduction* between fragments of mathematical practice.

---

289 Stephen Simpson, 'Friedman's Research on Subsystems of Second Order Arithmetic', in L. Harrington et al., eds., *Harvey Friedman's Research in the Foundations of Mathematics*, (Amsterdam: North-Holland, 1985), 137–159.

In that movement of proofs, Simpson detects certain sub-systems of second-order arithmetic that are '*canonical*' or '*archetypical*' enough to organize collections of theorems, and with them obtains a *natural stratification* of ordinary mathematics, where, for example, statements of type (a_0) the Heine-Borel Lemma, (a_1) the Bolzano-Weierstrass Theorem, or of type (b_0) the existence of prime ideals in rings, (b_1) the existence of maximal ideals in rings, can be classified in a precise hierarchy of equideductive complexity (in this case, $a_i \Leftrightarrow b_i$ $i=1,2$, in the eyes of a minimal constructive system).[290]

In the language of second-order arithmetic,[291] Simpson defines certain canonical subsystems of arithmetic in the following manner: RCA_0 incorporates the basic axioms of arithmetical terms, the axiom of induction $\varphi(0) \wedge \forall x(\varphi(x) \to \varphi(x+1)) \to \forall x \varphi(x)$ restricted to Σ_0^1 formulas and the comprehension axiom $\exists X \forall x(\varphi(x) \leftrightarrow x \in X)$ restricted to Δ_0^1 formulas (formulas in the Kleene-Mostowski

---

290 In what follows, we will define that minimal base, which we will denote RCA_0. The important point that should be emphasized here, before getting into the details, is that of *equideduction in the eyes of a weak axiomatic base*. Of course, from the absolute, analytic point of view of ZF set theory, we also have $ZF \vdash a_i \Leftrightarrow b_i$, simply because both statements are *tautological* from the absolute point of view of the axioms of set theory; but, in this case, the equideduction is *trivialized* and loses the logical richness of an intermediate derivation, without strong premises that distort it.

291 The language of second-order arithmetic extends that of the first order with variables of *two* types ('set' or second-order variables, in addition to 'numerical' or first-order variables), with an additional relational symbol ($\in$), with new formulas of the type $t \in X$ (where t is a numerical term, and X a set variable), and with additional quantification over set variables (whence the 'second' order). A formula in the second-order language is called *arithmetical* if it does not quantify over any set variables.

hierarchy);[292] WKL_0 consists of RCA_0 + a weak version of 'König's lemma' (every infinite subtree of a binary tree possesses an infinite branch); ACA_0 incorporates the basis axioms for arithmetical terms, the axiom of induction restricted to *arithmetical formulas* and the comprehension axiom, likewise restricted to arithmetical formulas. Other major subsystems of greater expressive power allow us to codify more sophisticated infinitary combinatorial manipulations, but we will not mention them here.[293]

With those nuclei ('archetypes') of deductive power, Simpson thus obtains a *real stratification* of the theorems of ordinary mathematics, demonstrating deep, logical equiconsistencies such as the following: in the eyes of WKL_0, a full equivalence (deduction and retroduction) between the system WKL_0 and many significant statements that can be demonstrated *within* WKL_0 (the Heine-Borel lemma, the completeness of first-order logic, the existence of prime ideals in countable commutative rings, the Riemann-integrability of continuous functions, local existence theorems for differential equations, the Hahn-Banach Theorem for separable spaces, etc.); in the eyes of RCA_0, a full equivalence between the system

---

292 Σ_n^0 includes (first-order) formulas with recursive matrices, which can be put into a normal form with n alternations of quantifiers and with $\exists$ as its outermost quantifier. Π_n^0 is defined likewise, for normal forms with $\forall$ as their outermost quantifier. We then define Δ_n^0 formulas as those that can be put into either $\Sigma_n^0 \cap \Pi_n^0$ form.

293 See Simpson, *Subsystems of Second Order Arithmetic*, chapters 5 and 6, 167–241 (transfinite recursion for arithmetical formulas and comprehension for Π_1^1 formulas).

ACA_0 and many of its theorems (the Bolzano-Weierstrass Theorem, the existence of maximal ideals in countable commutative rings, the existence of bases for vectorial spaces over countable fields, the existence of divisible closures for countable abelian groups, etc.). The usual deductive transit, which goes from the global (the system in its entirety) to the local (a particular theorem of the system), is here *reversed in a striking manner*, *allowing for an unexpected passage from the local to the global*, thanks to the theorem's equiconsistency with the entire system. It is instructive to observe how that reverse mathematics can emerge *only after* one makes explicit the natural candidates to act as *archetypical deductive nuclei* in second-order arithmetic.

Simpson describes one part of reverse mathematics as a restructuring of Hilbert's program:[294] in showing what the *minimal* axiomatic systems for proving the theorems of ordinary mathematics are, we can measure the latter's complexity in a precise fashion, and, in many cases, *reduce* that complexity to strictly finitary arguments. In this manner, although the *absolute* version of Hilbert's program fell with Gödel's Incompleteness Theorem, the program can nevertheless be *relativized*, and go on to secure certain partial and intermediary results. This, for example, is the case with Π_0^2 sentences provable

---

294 Ibid., 381–2.

in WKL_0: Simpson has proven a conservativity result[295] for WKL_0 over Peano arithmetic for Σ_1^1 formulas, which is in turn conservative over recursive arithmetic, and so the Π_0^2 sentences provable in WKL_0 can be proved with purely finitary arguments. In particular, this yields finitary proofs for many sophisticated mathematical results (expressible by Π_0^2 sentences provable in WKL_0): the existence of functional extensions (Hahn-Banach), of solutions to differential equations, of prime ideals in commutative rings, of algebraic closures, etc.

Certain archetypes in the spectrum of proofs in second-order arithmetic thus acquire a special relevance, since they end up being fully *reflected* in avatars of ordinary mathematics (abstract algebra, point topology, functional analysis, differential equations, etc.). As Simpson has pointed out, the archetypical nuclei of proofs emerge through a '*series of case studies*' leading to the discovery of 'the appropriate axioms',[296] showing how it is always mathematical *experimentation* that helps us declutter the landscape and find (when they exist) suitable invariants

---

295 Given a theory T in a language L, and another theory T_1 in a language $L_1 \supseteq L$, T_1 is a conservative extension of T if, for every L-sentence φ, $T_1 \vdash \varphi$ if and only if $T \vdash \varphi$. Simpson demonstrates that ACA_0 is a conservative extension of Peano arithmetic (first-order PA), and that both RCA_0 and WKL_0 are conservative extensions of PA restricted to Σ_0^1 formulas (in other words, $\Sigma_0^1 - PA$ is a first-order fragment of RCA_0 and WKL_0, while PA is a first-order fragment of ACA_0). The conservativity proofs exhibited by Simpson employ existential techniques from model theory, and their effective (finitary) character has been brought into question. Harvey Friedman has nevertheless stated that, in reverse mathematics, existential proofs of conservativity can indeed be converted into effective proofs.

296 Ibid., vii (xiii in 2nd edition).

lying behind the movement. The *processes* of mathematics (and *no longer* just its objects) are thereby forged in multiple webs of contradistinction, whether on the level of their logical representations, tied to well-defined archetypes of proof, or on the level of their structural correlations, tied to great spectra of regularity/singularity in domains of transit/obstruction. We once again find ourselves before an *abstract, relative differential and integral calculus*, like the one we have encountered in Grothendieck's work, and which, in the realm of reverse mathematics, consists in subtly *differentiating* the scales of demonstration, so as to afterward *reintegrate them* around canonical nuclei of proof.

The works of **Boris Zilber** (Russia, b. 1949) take the search for canonical structures for logic, algebra and geometry to a far greater depth.[297] His works install themselves in a new paradigm, progressively emerging in model theory, which shifts, in the middle of the twentieth century, from being understood as 'logic + universal algebra' (Tarski, Birkhoff; a paradigm normalized by Chang and Keisler),[298] to being considered, by the century's end, as 'algebraic geometry – fields' (Shelah, Hrushovski,

---

297 We are grateful to Andrés Villaveces for his lessons on Zilber (through an article we have referred to, conversations and conferences). An excellent overview of Zilber and his epoch can be found in B. Poizat, 'Autour du théorème de Morley' (the section '1980–90: les années-Zilber', in particular), in J.-P. Pier, ed., *Development of Mathematics 1950–2000* (Boston: Birkhäuser, 2000), 879–96.

298 H.J. Keisler & C.C. Chang, *Model Theory* (Amsterdam: North-Holland, 1973).

Zilber; a paradigm standardized by Hodges).[299] At issue here is an important change of perspective that brings model theory together with Grothendieck's visions (logical invariants of dimension that approach geometrico-algebraic invariants, o-minimal structures that approach 'tame' structures). Above all, this perspective situates logic in a *series of progressive geometrical decantations* of the objects and processes of mathematics. Along these lines, a revealing work[300] of Zilber's shows how certain strongly minimal theories[301] can be intrinsically associated with *combinatorial geometries*: the models of those theories are obtained as 'limits' of suitable finite structures, and the intrinsic geometries of those models, which are perfectly controlled, are either trivial (the algebraics do not expand to the model) or else affine or projective geometries over a finite field, something which allows optimal 'coordinate systems' for the models to be found. We see here how, behind general logical properties (strong minimality,

---

299 W. Hodges, *Model Theory* (Cambridge: Cambridge University Press, 1993). A contemporary version of model theory, including Zilber's contributions, appears in B. Poizat, *A Course in Model Theory: An Introduction to Contemporary Logic* (New York: Springer, 2000).

300 B. Zilber, 'Totally categorical theories, structural properties, and the non-finite axiomatizability', in L. Pacholski et al., *Proceedings of the Conference on Applications of Logic to Algebra and Arithmetic (Karpacz, 1979)* (Berlin: Springer, 1980), 380–410.

301 A theory is *strongly minimal* if the subsets definable in the models of the theory are finite or cofinite (for example, the theory of vector spaces, the theory of algebraically closed fields of characteristic p). This situation gives rise to natural logical notions analogous to those of *dimension* and algebraic closure, thereby opening the way to a reading of aspects of logic as fragments of a generalized algebraic geometry. The first result of Zilber's mentioned here refers to ω-categorical, strongly minimal theories (countable isomorphic models).

ω-categoricity), lie *deep geometrical nuclei*. If, following Grothendieck, we take up the terminological-conceptual tension between discovery and creation, we are thus confronted with archetypical *synthetico-geometrical* structures that are 'discovered' through the 'invention' of *analytico-logical* languages.

Another fundamental work[302] of Zilber's aims to classify the underlying geometries in strongly minimal theories. The *Zilber trichotomy* conjectures that strongly minimal theories can be split into three classes, according to their associated geometry: (*i*) theories with 'dismembered' models, which therefore possess a set-theoretic notion of dimension $(dim(X \cup Y) = dim(X) + dim(Y))$, and which cannot be interpreted in the theory of an infinite group; (*ii*) theories with basically linear models, whose geometry is *modular*, and therefore possesses a vectorial notion of dimension $(dim(X \cup Y) = dim(X) + dim(Y) - dim(X \cap Y))$, with finite fragments of the model being open to interpretation through abelian groups; (*iii*) theories with models that are bi-interpretable with the algebraically closed field of complex numbers (a particularly rich group), in which case the geometry is *algebraic*, and therefore has a natural notion of algebraic dimension. As Poizat has pointed out, among the great riches of the

---

302 B. Zilber, 'The structure of models of uncountably categorical theories', in *Proceedings of the International Congress of Mathematicians (Varsovia 1983)* (Varsovia: Polish Scientific Publications, 1984), 359–68.

trichotomy, as envisioned by Zilber, is the emergence of 'groups everywhere' – invisible at first, but lying in the depths ('archetypes').[303] In a sort of renaissance of Klein's Erlangen Program (1872),[304] groups – and their associated geometries – thus help to classify the deep forms of logic, and we may once again reflect on how logic *can in no way precede* mathematics, as is sometimes presumed from analytic perspectives.

Zilber's *conjectural* trichotomy aims to elucidate certain geometrical nuclei behind logical descriptions. About ten years after the conjecture's appearance, Hrushovski succeeded in demonstrating that there is, at least, a fourth case that it does not cover:[305] by means of a sophisticated *amalgamation* in the limits of models, Hrushovski constructed a strongly minimal structure whose geometry is neither trivial, nor modular, nor 'algebraic'.[306] Nevertheless, Zilber and Hrushovski conjectured and demonstrated that the trichotomy *is indeed valid*[307] for theories whose intrinsic geometries are *Zariski geometries*.[308] Afterward, by means

---

303 Poizat, 'Autour du théorème de Morley', 890.

304 For an illustrated modern edition of Klein's text, with a preface by Dieudonné, see F. Klein, *Le programme d'Erlangen* (Paris: Gauthier-Villars, 1974).

305 E. Hrushovski, 'A new strongly minimal set', *Annals of Pure and Applied Logic* 62, 1993: 147–66.

306 That is, not bi-interpretable with an algebraically closed field, in the sense of algebraic geometry.

307 E. Hrushovski & B. Zilber, 'Zariski Geometries', *Journal of the American Mathematical Society* 9, 1996: 1-56.

308 A Zariski geometry is a sort of variable topological structure $(X_n : n \geq 1)$, with no ethericity conditions and coherence between the X_n. Zariski geometries can be seen as *Lautmanian*

of a very subtle analysis of Hrushovski's counterexample, Zilber was led to conjecture[309] a new *extended alternative*: the intrinsic geometry of a strongly minimal theory must be (*i*) trivial, (*ii*) modular, (*iii*) algebraic (bi-interpretable with $(C, +, ., 0, 1)$, or (*iv*) 'pseudo-analytic' (bi-interpretable with an *expansion* of $(C, +, ., 0, 1)$ with an analytic function of the type *exp*).

The fundamental point here is the surprising appearance (again hidden and profound) of the *complex exponential*, with which Zilber conjectured that the classification could be closed. A first pathway in that direction consists in exploring model-theoretic notions of pseudo-exponentials in infinitary logics whose nuclei can be located,[310] so as to then construct limits of structures with pseudo-exponentials that allow us to cover case (*iv*). Another, completely unexpected, path of exploration lies

---

mixtures between model theory, algebraic topology and algebraic geometry. Hrushovski, drawing on techniques emerging in Zariski geometries, succeeded in demonstrating a Mordell-Lang conjecture regarding the number of rational points over curves in fields of functions (1996), which generalized the famous Mordell conjecture for curves over **Q** (proven by Faltings, winning him the Fields Medal in 1986). This is, perhaps, the most famous example of techniques *coming from logic* that help to resolve a problem in the 'heart' of mathematics. (Remember Connes.)

309 B. Zilber, 'Pseudo-exponentiation on algebraically closed fields of characteristic zero', *Annals Pure and Applied Logic* 132, 2004: 67–95.

310 The nucleus of the complex exponential $exp(2i\pi x)$ contains **Z**, where, with sum and multiplication, Peano arithmetic can be reconstructed, which gives rise to the many phenomena of incompleteness, instability, profusion of nonstandard models, etc. Nevertheless, using a logic with countable disjunctions $L_{\omega_1\omega}$, the nucleus of the exponential can be *forced* to be standard, by means of the sentence $\exists a \forall x \, (exp(x) = 0 \rightarrow \bigvee_{m \in Z} x = am)$ ($a = 2i\pi$ in the classical case), for example. The pseudo-exponentials generalize various properties of the complex exponential to arbitrary classes of models in $L_{\omega_1\omega}$, including the standard forcing of the nucleus.

in the new connections[311] that Zilber has found between pseudo-analyticity, foliations, and noncommutative geometry. Specifically, Zilber has 'discovered' that, on the one hand, the counterexamples to the trichotomy correspond to *deformations of Zariski curves by means of noncommutative groups*, such as 'quantum tori' groups, and that, on the other hand, other model-theoretic counterexamples are linked to certain foliations studied by Connes.[312] But, plunging even further into the depths,[313] into the abyssal model theory/noncommutative geometry hiatus, Zilber seems to be sensing the presence of the *common structures of physics*,[314] whose logical and geometrical

---

311 B. Zilber, 'Noncommutative geometry and new stable structures' (preprint 2005, available at http://www2.maths.ox.ac.uk/~zilber/publ.html).

312 Ibid., 2-3.

313 Wisdom lies sunken in the deep, in the infinite abysses, as Melville suggests in *Moby Dick*, in recounting Pip's second fall from the whaling vessel and his immersion in the ocean's lower strata: 'The sea had jeeringly kept his finite body up, but drowned the infinite of his soul. Not drowned entirely, though. Rather carried down alive to wondrous depths, where *strange shapes of the unwarped primal world* glided to and fro before his passive eyes; and the miser-merman, Wisdom, revealed his hoarded heaps', Herman Melville, *Moby Dick* (1849–51), H. Hayford, H. Parker & G.T. Tanselle), eds. (Evanston and Chicago: Northwestern University Press and The Newberry Library, 1988), 414. Archeal mathematics actively explores those 'strange shapes of the unwarped primal world' that escape the frightened Pip.

314 Regarding the junction that Zilber envisions between the objects of model theory and the deep structures of physics, Andrés Villaveces remarks, 'The structures that are most linked to noncommutative geometry and to the structures of physics are the *nonclassical Zariski geometries*. These form part of the "positive" side of the trichotomy, and appeared in the article by Hrushovski and Zilber. But only recently, only in the last two or three years, has Zilber begun to see that certain cases of "finite coverings" that should have been understood in terms of algebraic curves *cannot* be reduced to the latter. This dramatically changes two things: the junction with physics (more closely linked to a refined analysis of Zariski geometries – of finite but not unitary coverings of varieties that can only be understood in terms of actions of noncommutative groups), and the still more open role of pseudo-analytic structures' (personal communication, 2007).

representations would only be different facets of deep, unitary phenomena.[315]

The work of **Mikhael Gromov** (Russia, b. 1943) confirms, by way of other, utterly different routes, the richness of certain 'archetypical' connections revealed in contemporary mathematics. Considered perhaps the greatest geometer of recent decades, Gromov has completely revolutionized the various fields of investigation he has entered: geometry, where he introduced new perspectives on 'smoothing' and 'globalization' that are tied to the notion of *metrics everywhere*; partial differential equations, where he introduced homotopic calculations through *partial differential relations* (PDE via PDR); symplectic varieties, where he introduced into real analysis *pseudoholomorphic curves*, and, with them, new techniques of complex analysis; groups, where he introduced notions of polynomial growth, asymptotic behavior and *hyperbolicity*. In all of these fields, Gromov's works *combine* many of

---

315 The *great Russian school* – as we have seen with Drinfeld, Kontsevich and Zilber, and as we will soon see with Gromov – consistently tends to reveal deep unitary structures behind multiple mathematical and physical phenomena. This is also the case with the works of Vladimir Voevodsky (2002 Fields medalist), who has succeeded in providing a technical support for Grothendieck's motifs, as a *central trunk of cohomologies*. By introducing new Grothendieck topologies for algebraic objects, Voevodsky managed to construct subtle forms of 'surgery' for algebraic varieties – analogous to 'surgeries' for topological spaces, but having to overcome far more delicate obstructions – and succeeded in defining homotopic theories for algebraic varieties and schemes (2000). At the intersection between algebraic geometry and algebraic topology, Voevodsky's *eidal ascent* toward Grothendieck topologies would later allow him to bring about a *quiddital descent* toward the singular cohomology ('enriching it' in Voevodsky's sense) and pinpoint, in the last instance, the *archeal motifs* sought by Grothendieck. For technical introduction to Voevodsky's works, see Christophe Soulé, 'The work of Vladimir Voevodsky', in Atiyah, Iagolnitzner, *Fields Medalists' Lectures*, 769–72.

the qualities that Tao enumerates for what 'good mathematics' should be:[316] programmatic breadth of vision, conceptual inventiveness, technical mastery, abstract treatment, calculative skill, breadth of the spectrum of examples, deep interlacing of the global/abstract with the local/calculative, usefulness and applicability. The influence of the Russian school[317] is particularly palpable in this fantastic junction of geometrical vision, analytic virtuosity, and physical applicability.

To a large extent, Gromov's ideas surface in a complex counterpoint between refined webs of *inequalities* and series of suitable *invariants* in those webs. This is how it is with the triangular,[318] isoperimetric,[319] and

---

316 T. Tao, 'What is good mathematics?', preprint, arXiv:math.HO/0702396v1 February 13, 2007. Gromov's works attain a level of excellence in the majority of the qualifications that Tao specifies for 'good' mathematical work: the resolution of problems, technique, theory, perspicacity, discovery, application, exposition, vision, good taste, beauty, elegance, creativity, usefulness, power, depth, and intuition. Tao emphatically presents his list 'in no particular order' (ibid., 1) and, above all, makes an effort to illustrate the *correlationality* of certain of those qualities in concrete works of higher mathematics. And so, for the mathematician (Tao, for example), the *synthetic configuration of good qualities, in their coherent agglutination*, matters more than placing those qualities on a well-ordered, well-founded analytic scale.

317 Gromov completed his doctorate in 1969, at the University of Leningrad, under Rochlin. Concerning the influence of his soviet colleagues, see R. Langevin, 'Interview: Mikhael Gromov', in Pier, *Development of Mathematics*, 1,213–27 (1,221 in particular). Between 1974 and 1981, Gromov was a professor at Stony Brook; since 1981, he has been a permanent professor at the IHES. Profiting from Connes, Gromov and Kontsevich (among others), the IHES has been able to perpetuate the great tradition of higher mathematics opened by Grothendieck.

318 M. Berger, 'Rencontres avec un géomètre' (1998), in J.-M. Kantor, ed., *Où en sont les mathématiques* (Paris: Veuibert/Société Mathématique de France, 2002), 399–440. Berger's text emphasizes the inequalities mentioned here (400).

319 M. Gromov, *Metric Structures for Riemannian and Non-Riemannian Spaces* (Boston: Birkhäuser, 1999). Notes from a course given by Gromov at Paris VII, 1979–80. First published in French as M. Gromov, J. LaFontaine, P. Pansu, *Structures métriques pour*

systolic[320] inequalities, and how it is with quite diverse archeal constructs such as simplicial[321] and minimal[322] volumes, L^2-invariants, homotopic invariants linked to the geometry of partial differential relations, the Gromov-Witten invariants, etc. From a philosophical point of view, the emergence of these last few invariants is of particular interest. Over a given symplectic manifold, we can define many quasi-complex structures[323] that do not necessarily correspond to a complex manifold; in an effort to nevertheless study the symplectic/real with techniques from complex analysis, Gromov manages to overcome the *obstruction* by introducing a new notion of *pseudoholomorphic*

---

les variétés riemanniennes (Paris: Cedic-Nathan, 1981); the English edition contains extensive complements and appendices.) The sixth chapter, 'Isoperimetric inequalities and amenability' (321–49), presents a detailed study of various forms of isoperimetric inequality, by which the volume of a compact subspace is determined by means of the volume of its border. According to Berger ('Rencontres…', 415), isoperimetrics in infinite dimensions should be seen as a form of geometric surgery.

320 Sections 4E+ ('Unstable systolic inequalities and filling') and 4F+ ('Finer inequalities and systoles of universal spaces') (264–272) take up this theme directly. 'Systoles' are minimal volumes of nonhomologous cycles in the Riemannian manifold; a particular case consists in minimal noncontractible curves.

321 Given a compact manifold, its *simplicial volume* is defined as the infimum (greatest lower bound) of the sums of the (real) coefficients such that the fundamental class of the manifold is covered by the sums of those coefficients multiplied by simplicial sets. The simplicial volume turns out to be an invariant linked to the geometry of the manifold *in the infinite*, and it is therefore useful to study the varieties' *asymptotic* properties (Berger, 'Rencontres…', 412).

322 Given a compact manifold, its *minimal volume* is defined in the class of *all* the Riemannian structures linked to the manifold, through the metric that least raises the local behavior of protuberances in the manifold (Berger, 'Rencontres…', 413).

323 Given a manifold M, a quasi-complex structure over M is a section J of the fibration $End(TM)$ (TM being a tangent space) such that $J^2 = -Id$. If the manifold M is a complex manifold, then multiplication by i defines such a structure. For details, see G. Elek, 'The Mathematics of Misha Gromov', *Acta Mathematica Hungarica* 113, 2006: 171–85. Elek's article – prepared for the occasion of the awarding of the Bolyai Prize to Gromov in 2005 (a prize previously received only by Poincaré, Hilbert and Shelah!) – constitutes an excellent technical presentation of Gromov's work.

curve, which behaves magnificently well in the n-dimensional complex plane (any two points whatsoever can be connected by means of an appropriate pseudo-holomorphic curve); moving on to search for *invariants* of those curves, Gromov shows that the spaces *modulo* the curves are compact, and that it is therefore possible to work out a natural theory of homology, which leads to the Gromov-Witten invariants; in the last instance, the new invariants allow us, on the one hand, to distinguish an entire series of hitherto unclassifiable symplectic varieties, and, on the other, help to model unexpected aspects of string theory.[324]

In this manner, a direction of transit (real-complex), an obstruction in that transit (the multiplicity of the pseudo-complex), a partial saturation of the obstruction (pseudo-holomorphic curves), and an archeal deepening behind the new saturating concept (the Gromov-Witten invariants), show that mathematics – far from striving toward an analytical 'flattening' of phenomena's contradictory oscillations – *needs that deeply fissured topography* for its full development. In fact, in a brilliant analysis of situations of this sort, Gromov has pointed out[325] that

---

324 In the *A* model of string theory, 6 temporal dimensions are united in a 3-dimensional symplectic manifold, and the 'leaves of the universe' are parameterized as pseudoholomorphic curves over that manifold. The Gromov-Witten invariants are thus linked to deep physical problems. The interlacing of higher mathematics and cosmology is once again underwritten in unexpected ways.

325 Langevin, 'Interview: Mikhael Gromov', 1213–15.

'Hilbert's tree' (the ensemble of mathematics' branches), far from being simply planar and deductive, is crisscrossed by multidimensional geometrical objects: *exponential nodes* (sites in the tree bearing great, amplified oscillations: number, space, symmetry, infinity, etc.), *clouds* ('guides', or coherent gluings, such as geometrical nuclei à la Zilber, inside a tree that is a priori disconnected for reasons of complexity or undecidability), *local wells* (sites where mathematical information 'sinks' and is lost), etc.

Gromov's geometrical *style* recalls (implicitly) two of Grothendieck's *synthetic* strategies – a global vision of classes of structures and observation of properties on a large scale – and supplements them with an incisive, comparative *analytic*, through a *double fragmentation and reintegration* of the webs of inequalities under investigation. Gromov, in fact, offers a new understanding of Riemannian geometry by contemplating the class of *all* Riemannian manifolds and working with *multiple* metrics within that class – thereby setting the manifolds into motion and finding in that movement the appropriate relative invariants. Likewise, his works in the realm of partial differential relations inscribe themselves within a *double matrix* that allows surprising gluings to be performed along two primordial axes: synthetic/analytic and global/local. The h-principle[326] (h for homotopy)

---

326 M. Gromov, *Partial Differential Relations* (New York: Springer, 1986).

effectively postulates, in certain geometrical realms, the existence of certain homotopic deformations between continuous sections of a sheaf (linked to local differential correlations that codify the local conditions in a partial differential equation) and the sheaf's *holonomic* sections (tied to global solutions, through global differentials). The monumental work undertaken by Gromov in his *Partial Differential Relations* succeeds in exhibiting the *ubiquity* of the *h*-principle in the most remote areas of geometry (the synthetic richness of the principle) and constructing a multitude of local methods and practices for verifying the *h*-principle in particular conditions (analytic richness).

The group structure, rejuvenated by the likes of Connes and Zilber, seems to enjoy infinite lives in Gromov's hands. His program in *geometrical group theory* can be described as an intent to characterize finitely generated groups, modulo quasi-isometries, which is to say, modulo 'infinitesimal' deformations of Lipschitz-type distances.[327] In that program, Gromov has demonstrated that many properties of groups turn out to be quasi-isometric *invariants*; in particular, the (word-)*hyperbolicity* of a group[328] is one such invariant, by which we can characterize the *linear*

---

327 For technical details, see Elek, 'The Mathematics of Misha Gromov', 181–2.

328 The basic example of a *hyperbolic group*, in Gromov's sense, is the fundamental group of an arbitrary manifold of negative curvature. The generalization of certain properties of 'thin' triangles in the universal covering of that variety leads to abstract definitions of hyperbolicity (ibid., 183).

complexity of a group's associated word problem.[329] On the other hand, using the definition of a *metric* in a Cayley graph of a finitely generated group, Gromov is able to define the 'polynomial growth' of a group and study that asymptotic growth's correlations with classical properties: solubility, nilpotence, Lie sub-representations, etc.[330]

We thus find ourselves before a quite typical situation in contemporary mathematics, where *certain classical nuclei are seen as the limits of deformations* (be they logical, algebraic, topological or quantic) in very broad classes of spaces. In virtue of these great synthetic processes (in the style of Zilber's 'groups everywhere', or Gromov's 'metrics everywhere'), the classical invariants are recuperated, but many new invariants (*archeal* in our terminology) are discovered as well, invariants that a restricted vision – whether local, analytic or classical – *cannot* catch sight of.

---

329 Given a recursively presented group G, the *word problem* associated with G consists in deciding if two finite products of the generators of G (that is to say, words in the free group) coincide or not. Some groups for which the word problem is *soluble* include finite groups and simple, finitely generated groups. It can be demonstrated that a uniform solution of the problem for all groups does not exist, and so the *measure* of the problem's complexity for certain classes of groups turns out to be a result of great interest. Gromov has demonstrated that the complexity of the word problem for a given group is linear if and only if the group is hyperbolic. By means of a *metric* in the class of finitely generated groups, it has been demonstrated that the *closure* of the subclass of hyperbolic groups contains the 'Tarski Monsters'. (The latter are infinite groups whose nontrivial subgroups are cyclical groups of order p, for a fixed prime p; the existence of such groups was proven by Olshanskii [1980], with $p > 10^{75}$ – a result that should make philosophers dream!)

330 Ibid., 183–4.

PART THREE

Synthetic Sketches

CHAPTER 8

FRAGMENTS OF A TRANSITORY ONTOLOGY

In this, the third part of the work, we will reflect (philosophically, methodologically and culturally) on the case studies that we have presented in the second part. Accordingly, *when we refer to 'mathematics' (and its derivative adjectives) in what follows, we mean 'contemporary mathematics'*, unless we explicitly state otherwise. Now, we should note right away that this essay cannot cover all the forms of doing mathematics, and, in particular, will not dwell on the practices peculiar to elementary mathematics. We therefore do not aim to produce anything like an all-encompassing philosophy of mathematics, only *to call attention to a very broad mathematical spectrum that has rarely been accounted for in philosophical discussions*, and which should no longer be neglected. In the final chapter we will try to provide an intrinsic characterization of the interval between 1950–2000 (open on both extremes) with regard to 'contemporary mathematics'; but for the moment, we we will merely ground ourselves on the concrete cases of mathematical practice reviewed in part 2. We will provide an extensive number of cross references to those case studies; to that end, *we will systematically use references between square parentheses, such as [x], [x–y] and* [*x, y, ...*] *in the body of the text*, which will direct the reader to the pages *x*, *x–y*, or *x*, *y*, ... of this book.

The case studies of the second part should have made it clear that contemporary mathematics is incessantly occupied with *processes of transit* in exact thought, involved in multiple *webs of contradistinction*, both internal and external. From this it immediately follows that the questions concerning the content and place of mathematical objects – the ontological 'what' and 'where' – through which we hope to describe and situate those objects, cannot be given absolute answers, and *cannot be fixed* in advance. The *relativity* of the 'what' and the 'where' are indispensable to contemporary mathematics, where everything tends toward transformation and flux. In this sense, the great paradigm of Grothendieck's work, with its profound conception of a relative mathematics [140–141] interspersed with changes of base of every sort in very general topoi [141–142], should be fully understood as an 'Einsteinian turn' in mathematics. As we have seen, we are dealing with a vision that ramifies through all the mathematics of the epoch, and which is also capable of giving rise to a genuine *Einsteinian turn in the philosophy of mathematics*.

Now, the point of Einstein's theory of relativity, once we assume the movement of the observers, consists in finding suitable invariants (no longer euclidean or Galilean) behind that movement. Likewise, the point of a relative mathematics à la Grothendieck, once we assume the transit of mathematical objects, consists in finding suitable invariants (*no longer elementary or classical*) behind that transit. This is the case with many of the archeal situations in mathematics that we have been looking at: motifs [144–146], PCF theory [201–202], intermediate allegories [245–246], Zilber's extended alternative [256], the *h*-principle [263–4], etc. A skeptical relativism, which leads to disorientation and allows for an isotropy of values, in the style of certain postmodern subrelativisms or the infamous *pensiero debole*, is thus very far from being the same as the Einsteinian or Grothendieckian projects, where, though there be neither absolute foundations nor fixed objects, not everything turns out to be comparable or equivalent, and where we can calculate *correlative archeal structures* – that is, invariants with respect to a given context and a given series of correlations – which, precisely, *allow differences to be detected and reintegrated*.

The first important point in specifying 'what' mathematical objects are consists in really taking relativity and transit in contemporary mathematics *seriously*. Objects in this realm cease to be fixed, stable, classical and well founded – in sum, they cease to be 'ones'. Instead they tend toward the mobile, the unstable, the nonclassical, and the merely contextually founded – in short, they approach 'the many'. *Multiplicity* everywhere underlies contemporary transit, and the objects of mathematics basically become *webs and processes*. Determinate 'entities', firmly situated in *one* absolute, hard and fast universe, do not exist; instead we have *complex signic webs* interlaced with one another in *various* relative, plastic and fluid universes. The levels contemplated by these 'complex signic webs', where mathematical objects are constituted, are multitudinous, and *no fixed level exhausts the richness of the object* (*web*).

This is obvious, for instance, with the mathematical 'object', *group*; we have seen how that object appears and captures disparate information (under the most diverse representation theorems) in the most distant realms of mathematics: homology and cohomology groups **[142–148, 178–179]**, Galois groups **[150, 155, 225]**, group actions **[162–163, 180–181]**, abelian groups **[165]**, homotopy groups **[176]**, algebraic groups **[184]**, the Grothendieck-Teichmüller group **[225, 233]**, Lie groups **[223]**, quantum groups **[223]**, Zilber groups **[255–256]**, hyperbolic groups **[264]**, etc. It is not that we are confronted here, ontologically, with a universal structure, that takes on supplementary properties at each supposed level of reading (logical, algebraic, topological, differential, etc.); what happens is rather that the diverse webs of mathematical information codified under the structure of group *overlap* ('*presynthesis*') *and compose* ('*synthesis*') *so as to transmit information in a coherent fashion*. It is not that there exists 'one' fixed mathematical object that could be brought to life independently of the others, in a supposedly primordial universe; what we find instead is the plural existence of *webs incessantly evolving as they connect* with new universes of mathematical interpretation. This is particularly evident in the inequality webs **[260]** studied by Gromov, or in the webs of equideducible theorems **[248–249]** that Simpson has displayed; the progressive leaps and bounds taking place in the webs go on to configure the global landscape, and this *modifies in turn* the entities that the global environment locally internalizes.

Given that the objects of mathematics 'are' not *stable sums* but *reintegrations of relative differentials*, the question concerning their situation ('where they live') takes on an aspect that is almost *orthogonal* to the way in which this question is posed from an analytic perspective (grounded in set theory). For if mathematics finds itself in perpetual transit and evolution, the situation of an object cannot be anything but relative, with respect to a certain realm ('geography') and to a moment of that realm's evolution ('history'). This just goes to reinforce the position of Cavaillès, who understood mathematics as gesture – a position that has echoed throughout the century, all the way to Gromov, who pointed out how we should come to 'admit the influence of historical and sociological factors'[331] in the evolution of Hilbert's tree **[262]**.

---

331 Langevin, 'Interview: Mikhael Gromov', 1,214.

Needless to say, this reading of mathematics as a *histori-cal* science goes against the grain of the readings offered by the analytic philosophy of mathematics – readings according to which fragments of the edifice arise atempo-rally against *absolute backgrounds*, codified in the various analytic 'isms' **[102–103]**, and within which each com-mentator plays at undermining contrary positions, and proposing his version as the most 'adequate', which is to say, as the one most capable of resolving the problematics at stake. Curiously, however, the supposed *local reconstruc-tions* of 'mathematics' – studied a hundred times over in the analytic texts – go *clearly against* what mathematical logic has uncovered in the period from 1950 to 2000. In fact, as we have seen, following the style of Tarski (logics as fragments of algebras and topologies) and Lindström (logics as coordinated systems of classes of models), the most eminent mathematical logicians of the last decades of the twentieth century (Shelah, Zilber, Hrushovski) have emphasized the emergence of deep and hidden geometrical kernels **[195, 253]** lying behind logical manipulations. Just as Jean Petitot, in his *mixed studies*[332] on neurogeometry, Riemannian varieties and sheaf logic (Petitot declares himself a great admirer of Lautman's

---

332 See J. Petitot, 'Vers une neuro-geometrie. Fibrations corticales, structures de contact et contours subjectifs modaux', *Mathématiques, Informatique et Sciences Humaines* 145, 1999: 5–101, or 'The neurogeometry of pinwheels as a subriemannian contact structure' *Journal of Physiology* 97, 2003: 265–309. His remarkable doctoral thesis (*Pour un schématisme de la structure* [EHESS 1982, 4 vols.], part of which is contained in *Morphogenèse du sens* [Paris: PUF, 1985]) has still not been made use of much in mathematical philosophy.

'mixtures'), has begun to defend the idea that geometrical proto-objects underlie neuronal activity, and, moreover, that a *proto-geometry should take heuristic precedence over a language*, contemporary mathematical logic has likewise come to demonstrate how *a proto-geometry necessarily precedes a logic*. What is at stake, then, is a situation that leads us to completely *overturn* – again, almost orthogonally – the usual approaches of analytic philosophy.

Within contemporary mathematics, and following the 'double orthogonality' that we have pointed out, an object is not something that 'is', but something that is *in the process of being*, and these occurrences are not situated in a logical warp, but in an initial spectrum of *proto-geometries*. The consequences for an ontology of mathematics are immense and radically innovative. On the one hand, from an internalist point of view, the 'what' involves webs and evolving transits, while the 'where' involves proto-geometrical structures anterior to logic itself. On the other hand, from an externalist point of view, the 'what' takes us back to the unexpected presence of those proto-geometries in the physical world (the actions of the Grothendieck-Teichmüller group on the universal constants of physics **[224–225,233]**, the Atiyah index **[208–209]**, the Lax-Phillips semigroup **[218–219]**, the interlacing of Zilber's pseudo-analyticity with the physical models of noncommutative geometry **[257–259]**, etc.), while the 'where' conceals a profound dialectic of relative correlations between concrete phenomena and their theoretical representations. In fact, what comes to light in these readings is that the questions concerning an absolute 'what' or 'where' – whose answers would supposedly describe or situate mathematical objects once and for all (whether in a world of 'ideas' or in a 'real' physical world, for example) – are *poorly posed questions*.

The richness of mathematics in general (and of contemporary mathematics in particular) consists precisely in *liberating* and *not restricting* the experience and influence of its objects. In a certain sense, a *base* from which we may better understand mathematics' indispensable *transitability* could be furnished by the mixture of (*i*) *structuralism*, (*ii*) *categories* and (*iii*) *modalization* that Hellman proposes **[106–107]**, but in reality the situation is more complex, as comes to light in the multiple oscillations, hierarchizations and ramifications cast in relief in this book's second part. In fact, if a categorical reading (*ii*) seems indispensable in contemporary mathematics (something we have emphasized with multiple case studies beneath which, implicitly or explicitly, lies the Grothendieckian categorical and relativistic program) and if, beyond a certain threshold of complexity, mathematical modalization (*iii*) seems equally indispensable (the *multiform transit* between hypotheses, models, calculations and contradistinctions, carefully hidden in the classical formalization of set theory, and carefully ignored by so many of the 'hard' currents of the analytic philosophy of mathematics), contemporary mathematics nevertheless underscores the importance of *processes* over structures (*i*), since the latter emerge on a *second level* as the invariants of suitable processes. The relative ontologization of objects in contemporary mathematics thus forces us to further *dynamize* Hellman's base (which appears fixed in a quadrant of Shapiro's square [10]), so that we can better take stock of the contemporary processes.

Ultimately, when faced with contemporary mathematics we cannot escape a certain 'transitory ontology' that, at first, terminologically speaking, *seems* self-contradictory. Nevertheless, though the Greek *ontotetês* sends us, through Latin translations, to a supposedly atemporal 'entity' or an 'essence' that 'ontology' would study,[333] there is no reason (besides tradition) to believe that those entities or essences should be absolute and not *asymptotic*, *governed by partial gluings in a correlative bimodal evolution* between the world and knowledge. The philosophical bases of such a 'transitory ontology' can be found in Merleau-Ponty's theory of *shifting*[334] (in the general realm of knowledge and the particular realm of visuality), and in the specific

---

333 J.-F. Courtine, 'Essence', in Cassin, *Vocabulaire européen des philosophies*, 400–14.

334 See M. Merleau-Ponty, *Notes des cours du Collège de France (1958–59, 1960–61)* (Paris: Gallimard, 1996), and *L'oeil et l'esprit* (Paris: Gallimard, 1964). The latter is the last text he published in his lifetime, in 1961, and is a magnificent way to be introduced to Merleau-Ponty's work. See also his two posthumous works, *La prose du monde* (Paris: Gallimard, 1969) and *Le visible et l'invisible* (Paris: Gallimard, 1964). The hiatus between the visible and the invisible can only be understood while shifting within it.

transitory ontology proposed by Badiou[335] (in the field of mathematics). For Merleau-Ponty, the 'height of reason'[336] consists in *feeling the shifting of the soil*, in detecting the movement of our beliefs and supposed claims of knowing: 'each creation changes, alters, clarifies, deepens, confirms, exalts, recreates or creates by anticipation all the others'.[337] A complex and mobile tissue of creation surges into view, full of 'detours, transgressions, slow encroachments and sudden drives', and in the contradictory coats of sediment emerges the force of creation entire. This is exactly what we have detected in contemporary mathematics, as we have presented it in part 2 of this work. For Badiou, mathematics is a 'a pseudo-being's quasi-thought', *distributed* in 'quasi-objects'[338] that reflect *strata* of knowledge and world, and whose *simple correlations* (harmonic and aesthetic) *grow* over time. Those *quasi*-objects not only escape fixed and determinate identities: they proceed to evolve and *distribute* themselves between warps of ideality and reality. Through the dozens of situations surveyed in part 2, we have indeed been able to make evident *the distribution*, *the grafting and gluing* of those quasi-objects, not only between fragments internal to mathematics, but between

---

335 Badiou, *Briefings on Existence*.

336 Merleau-Ponty, *L'Oeil et l'esprit*., 92.

337 Badiou, *Briefings on Existence*, 47.

338 Ibid., 42-3.

theoretical mathematics (the realm of the eidal) and the concrete physical world (the realm of the quiddital).

With new perspectives and with new force, Badiou's transitory ontology allows us to bring an *untrivialized Plato* back to the landscape of mathematical philosophy. In the style of Lautman's *dynamic Platonism* – attentive to the composition, hierarchization and evolution of the mixtures in the *Philebus*[339] – Badiou, too, insists on a Platonism essentially *open* to a 'cobelonging of the known and the knowing mind', through which one derives an 'ontological commensurability'[340] that incorporates movement and transit. We thereby immerse ourselves in a Platonism that is not static, not fixed to a supposedly transcendent world of Ideas, and very far from the image most often used to sum up the doctrine.[341] Badiou's dynamic Platonic reading opens onto quite different philosophical perspectives – in particular, onto an understanding of mathematics as evolving *thought*[342] and a study of the problems of *saturation, maximization, and invariance* in the   movements of

---

339 Lautman, *Essai sur l'unité...*, 143–7, 203, 227–8, 303. See N.-I. Boussolas, *L'Être et la composition des mixtes dans le Philèbe de Platon* (Paris: PUF, 1952).

340 Badiou, *Briefings on Existence*, 90.

341 This is what we find in the 'trivialization' presented by Benacerraf: 'Platonists will be those who consider mathematics to be the *discovery* of truths about structures that exist independently of the activity or thought of mathematicians'. In Benacerraf & Putnam, *Philosophy of Mathematics*, 18.

342 Badiou, *Briefings on Existence*, 39–54, 102. Mathematics as 'thought' studies the exact transitions of being and approaches a transformative/productive *gesturality* (Cavaillès, Châtelet) of *nontrivial* information. This position is quite far from an understanding of mathematics as a mere language game, or as a set of tautological deductive transitions.

thought – which, following Merleau-Ponty, falls within the *asymptotic order of 'shifting'*, and which covers some of the deeply mathematical problematics to which we have called attention in the case studies contained in part 2.

The *process of being*, *commensurably in between the webs of cognition and the webs of phenomena*, is a fundamental characteristic of 'transitory ontology' that contemporary mathematics *requires*. What we are dealing with here is an *apparently innocuous change of grammatical category* in posing ontological questions, which nevertheless affords them a whole new dimensionality and enriches the problematics at stake: the 'what' and the 'where' – which were initially taken up in an absolute and actual present, leading us to ask poorly posed questions – come to be understood in a *modal and relative present perfect*. The major consequence of this opening up to transit is to immediately shatter the dualisms and pigeonholes of Shapiro's square [10]. As we have seen, the objects of mathematics – that is, in reality, its processes and quasi-objects – *ceaselessly transit between eidal, quiddital and archeal webs*, whether in the world interior to mathematics, or in their perpetual contradistinction with the physical exterior. What in one context turns out to be eidal (Grothendieck-Teichmüller (GT), linked to the Euler numbers [225], for example), in another appears as quiddital (GT acting on the universal formulas of physics [233]), and in yet another appears as archeal (GT in the dessins d'enfant [225]).

The classical dissociations and exclusions (of the either-or variety, in the style of Benacerraf's dilemma [11]) have caused excessive and unnecessary damage to the philosophy of mathematics, and the time to overcome them is at hand. To that end, we can already count on at least three great tendencies in contemporary mathematics – where binary positionings are cast aside and where perspectives of continuity between diverse webs are opened up – that philosophical reflection should begin to take seriously: the understanding of the 'positive' (classical, commutative, linear, elementary, structured, etc.) as a *limit* of 'negative' mediations (intuitionism, noncommutativity, nonlinearity, nonelementarity, quantization, etc.); the theory of *sheaves*, with its continuous handling of coherence and gluing between the local and the global; the mathematical theory of *categories*, with its handling of differentiation and reintegration between the particular and the universal. In what follows, we will clarify these three deep technical tendencies from a conceptual and philosophical perspective, and explain how they allow us to *reinforce* a transitory ontology of mathematics.

Among the characteristics specific to mathematics between 1950 and 2000, we have indicated the act of working with the *fluxion* and *deformation* of structures' typical boundaries [42]. This is obvious, for example, in Grothendieck's K-theory [135–136, 210–212], where we study the perturbations of morphisms over classical fibers; in Shelah's abstract elementary classes [196–199], where we study limits of algebraic invariants beyond first-order classical logic; in the Lax-Phillips semigroup [218–219], where non-euclidean geometry allows us to understand the scattering of waves; in Connes's noncommutative geometry [222–224], where the dispersion and fluxion of quantum mechanics and thermodynamics are unveiled; in Kontsevich's quantizations [229–233], where classical structures obtain as limits of the quantized; in Freyd's allegories [243–246], where topoi are seen as limits of intermediate categories; in Zilber's extended alternative [256], where counterexamples to the trichotomy arise as deformations of noncommutative groups; or in Gromov's theory of large-scale groups [264–265], where we study the nonlinear asymptotic behavior of finitely generated groups. These are all examples of *high* mathematics, with *great conceptual and concrete content* (and are thus far from being reducible to mere 'language games'). In them, a new understanding of mathematical (quasi-)objects is demonstrated, by means of fluxions, deformations and limits, and by passing through intermediate strata that are nonpositive (nonclassical, noncommutative, nonlinear, nonelementary, and nonquantized).

A result due to Caicedo[343] shows that *classical logic in a 'generic' fiber of a sheaf of first-order structures is no more than an adequate limit of intuitionistic logic in the 'real' fibers of the sheaf.* This remarkable situation shows that the construction of the classical and the positive as 'limit idealizations', as seen in the aforementioned *mathematical* examples, is reflected in the realm of *logic* as well, and in exactly the same fashion. What comes to the surface, once again, is the *continuity of mathematical knowledge*, for which watertight compartments are worthless. The richness of Caicedo's sheaf logic is precisely due to its being elaborated in an *intermediate zone* between Kripke models[344] and Grothendieck topoi **[141–144]**, benefiting from the many concrete examples of the former and the abstract general concepts of the latter. In a *total crossing of algebraic, geometrical, topological and logical techniques*, Caicedo constructs an instrumentarium to incite transit

---

343 X. Caicedo, 'Lógica de los haces de estructuras', *Revista de la Academia Colombiana de Ciencias Exactas, Físicas y Naturales* XIX, 74, 1995: 569–85. Caicedo provides a framework of great depth and breadth – mathematical, logical, conceptual, and philosophical – which is unfortunately still unknown by the international community. He has announced (in 2012) a forthcoming publication in English. For a partial overview in Italian, see F. Zalamea, 'Ostruzioni e passaggi nella dialettica continuo/discreto: il caso dei grafi esistenziali e della logica dei fasci', *Dedalus. Rivista di Filosofia, Scienza e Cultura – Università di Milano* 2, 2007: 20–5.

344 *Kripke models* are 'trees' that can be used to represent a branching temporal evolution; from a mathematical point of view, they are simply *presheaves* over an ordered set (seen as a category). Kripke models furnish a complete semantics for *intuitionistic* logic. Other complete semantics for intuitionism are provided by the class of topological spaces, or by the class of elementary topoi **[11]**. Intuitionism and continuity are interlaced over an archeal ground in this manner, and Thom's aporia emerges in a new form: classical-discrete *versus* intuitionistic-continuous. This new appearance of intuitionism, untethered from its original constructivist aspect, has not been sufficiently exploited in the philosophy of mathematics.

and transference. The result is his 'fundamental' theorem of model theory, where pivotal statements in logic such as the Loz theorem for ultraproducts, the completeness theorem for first-order logic, forcing constructions in sets, and theorems of type omissions in fragments of infinitary logic, can all be seen, uniformly, as constructions of generic structures in appropriate sheaves.

From a philosophical point of view, a striking consequence follows from all of this. Through fluxions and transit we effectively see that *the classical perspectives are no more than idealities, which can be reconstructed as limits of nonclassical perspectives that are far more real*. A particular case of this situation is a new synthetic understanding of the point-neighborhood dialectic, where – contrary to the analytic and set-theoretic perspective, according to which neighborhoods are constructed from points – the classical, ideal 'points', which are *never seen* in nature, are constructed as limits of real, nonclassical neighborhoods, which, by contrast, are connected to *visible*, physical deformations. From this perspective, the ontology of the (quasi-)objects at stake once again undergoes a radical turn: an 'analytic' ontology, linked to the study of set-theoretic classes of points, can be no more than an *idealized fiction*, which *forgets* an underlying 'synthetic' / 'transitory' ontology that is far more real, and linked to the study of physically discernible, continuous neighborhoods. The transitory and continuous (quasi-)objects of nonclassical mathematics are thus *interlaced* with transitory and continuous phenomena in nature, through webs of informative correlation that are gradual, nonbinary and nondichotomous.

This situation is corroborated by the principal paradigms of *sheaf theory*.[345] The ancient philosophical problematic, 'how do we get from the Many to the One?' (corresponding to a phenomenological transit) becomes the mathematical problematic, 'how do we get from the local to the global?' (technical transit), which, in turn, can be subdivided into the questions (*i*) 'how do we differentially register the local?' and (*ii*) 'how do we globally integrate those registers?' The natural mathematical concepts of neighborhood, covering, coherence and gluing emerge in order to *analytically* tackle question (*i*),

---

345 Sheaves emerge in the work of Jean Leray, in the study of indices and coverings for differential equations (with his works in the Oflag XVII, and articles from 1946–50; the term 'sheaf' first appears in 1946). In the general realm of the study of a manifold through its projections into manifolds of lower dimensions (Picard, Lefschetz, Steenrod), there arises the problematic of studying the topology of the initial manifold by means of the coherent information provided by the fibers in the projection, and sheaves are precisely what help to capture (and glue together) the continuous *variation* in the fibers. The Cartan Seminar of the École Normale Supérieure (1948–51) served to distill Leray's ideas and present the sheaf concept as it is known today: as a fibered space, or 'étalé space' in Lazard's terminology (not to be confused with Grothendieck's notion of étale: distinguished by a mere *accent aigu*, these two concepts are almost diametrically opposed – the first is ramified, the second is *non*ramified), and as the sheaf of germs of sections. Godement unified the concepts and terminology in his *Topologie algébrique et théorie des fasceaux* (1958), in parallel with Grothendieck's *Tohoku* **[162–166]**. For a detailed history of sheaf theory, see J. Gray, 'Fragments of the history of sheaf theory', in *Lecture Notes in Mathematics* 753 (New York: Springer, 1979), 1–79, and C. Houzel, 'Histoire de la théorie des fasceaux', in *La géométrie algébrique* (Paris: Albert Blanchard, 2002), 293–304. In our final chapter we will situate the appearance of sheaf theory as the *concretization of a deep conce‚ptual break* between modern and contemporary mathematics, and we will try to show that the diachronic break around 1950 is not a mere historical accident.

while the natural mathematical concepts of restrictions, projections, preservations and sections emerge in order to *synthetically* tackle question (*ii*). Presheaves (a term we owe to Grothendieck) cover the combinatorics of the *discrete* interlacings between neighborhood/restriction and covering/projection, while sheaves cover the *continuous* combinatorics tied to the couplings of coherence-preservation and gluing-section (figure 13). In this way, the *general concept of sheaf* is capable of integrating a profound web of correlations in which aspects both analytic and synthetic, both local and global, and both discrete and continuous are all incorporated.

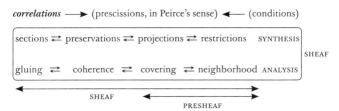

Figure 13. *The sheaf concept: general transitoricity of analysis/synthesis, local/global, and discrete/continuous.*

In *prescinding*[346] the fundamental notion of *gluing* in the sheaf, the notions of coherence, covering and

---

346 *Prescission*, in Peirce's sense, allows for the passage from the particular to the general, *descrying* the most abstract in the most concrete. It is a process that is ubiquitous in mathematics, and one that also takes place in the passage from the Many to the One, which is to say, in the phenomenological search for universal categories. Peirce's *three cenopythagorean categories* (firstness: immediacy; secondness: action-reaction; thirdness: mediation) surface in meticulous dialectics of prescission. For a detailed study of those

neighborhood emerge *progressively and necessarily*. These last two notions are of the greatest importance in ontological discussions. On the one hand, if it turns out that the 'real' cannot, as we have seen, be anchored in the absolute, and if the 'real' can therefore only be understood by way of *asymptotic conditions*, then strategies for *covering* fragments of the real take on a pivotal ontological importance. It is in this sense that we should recall Rota's exhortations that, with respect to mathematical objects, we must escape the 'comedy of existence' staged by analytic philosophy, and instead attend to a 'primacy of identity'[347] that is bound to the grafting of *the real's modal coverings*, and which will allow us to classify the possible identities between ideas and physical objects, identities that evolve – and in the most interesting cases remain invariant – over large-scale periods (here we should remember Gromov). On the other hand, a turn toward a *real logic of neighborhoods*, as a counterpoint to a logic that is ideal and punctual – a turn to *sheaf logic* as a counterpoint to classical logic – likewise leads to a radical ontological turn. Many of the exclusive disjunctions presupposed by analytic and classical thought are undermined in synthetic environments (whether physical or

---

dialectics – linked to 'bipolar tensions' similar to those that have been discovered in the *adjunctions* of mathematical category theory – see De Tienne, *L'analytique de la répresentation...*

347 Rota, *Indiscrete Thoughts*, 184–6.

cognitive) where mediations are the rule. In such synthetic/transitory/continuous realms, in particular, there is no reason why the disjunctive question concerning the 'ideality' *or* 'reality' of mathematical (quasi-)objects should be answered in an exclusive fashion [6–14], as is suggested by the watertight compartments of Shapiro's table [10]. In concrete terms, sheaves, *those (quasi-) objects indispensable to contemporary mathematics*, acquire all their richness in virtue of their *double* status as ideal/real, analytic/synthetic, local/global, discrete/continuous; the mutual inclusion (and not the exclusion)[348] of opposites – an incessant exercise of mediation – secures their technical, conceptual and philosophical force.

Category theory axiomatizes *regions* of mathematical practice, according to the structural similarities of the objects in play and the modes of transmission of information between those objects, in harmony with Peircean pragmaticism [111–121], which is equally sensitive to *problems of transference*. In an inversion of set theory, where objects are internally analyzed as a conglomerate of elements, category theory studies objects through their *external*, *synthetic* behavior, in virtue of the object's *relations* with its environment. A morphism is *universal* with

---

348 Rota clearly expresses the urgency of not adopting exclusions in an a priori fashion: 'Mathematical items can be viewed either as analytic statements derived within an axiomatic system or as facts about the natural world, on a par with the facts of any other science. Both claims are equally valid. [...] The contextual standing of an item as analytic or synthetic is not fixed'. Ibid., 168.

respect to a given property if its behavior with respect to other similar morphisms in the category possesses certain characteristics of unicity that distinguish it within the categorical environment. The basic notions of category theory associated with universality – the notions of *free object* and *adjunction* – respond to problematics bound up with the search for *relative archetypes and relative dialectics*. In the multiplicity, in the wide and variable spectrum of mathematics' regions, category theory manages to find certain patterns of universality that allow it to overtake the splintering of the local and the surmounting of concrete particulars. In a category, for example, a free object can be projected into *any* object of the category's sufficiently large subclasses: it is therefore something of a *primordial sign*, incarnated in all of the correlated contexts of interpretation. Beyond relative localizations, certain *relative universals* thus arise, giving an entirely new technical impulse to the classical notions of universality. Though we can no longer situate ourselves in a supposed absolute, nor believe in concepts that are universally stable in space and time, *the notions of universality have been redimensioned* by category theory, and coupled with a series of transferences relative to the universal-free-generic, facilitating a *transit* behind which are revealed remarkable *invariants*.

In this manner, category theory explores the structure of certain *generic entities* ('generalities') by a route akin to the 'scholastic realism' of the later Peirce. Categorical

thought contemplates a dialectic between universal definitions in abstract categories (generic morphisms) and
realizations of those universal definitions in concrete
categories (classes of structured sets), and within the
abstract categories we can behold *universals* that are *real*
but not existent (that is to say, that are not incarnated
in concrete categories: think, for example, of an initial
object, definable in abstract categories, but whose reality
has no incarnation in the category of infinite sets, where
initial objects do not exist). The transitory ontology of
mathematical (quasi-)objects therefore opens onto an
intermediate hierarchy of modes of universality and modes
of existence, beyond restrictive binary demarcations. In
particular, as Lawvere has pointed out, the objects of
elementary topoi (including presheaves and sheaves)
[195] should be seen as *variable sets*, with modes of
belonging that fluctuate over time. The ontological transitoriness of those 'entities' could not be more obvious.

In the spectrum of pure possibilities, the Peircean
pragmaticist maxim [111–121] must confront the
idea of concepts that are universal and logically correct,
but which cannot be adequately incarnated in restricted
contexts of existence. The thought of mathematical category theory recaptures this kind of situation with great
precision. In line with tendencies in universal algebra
and abstract model theory, for example, category theory
has been able to define very general notions of *relative*

universal semantics as appropriate invariants for given classes of logics. Behind the multiplication of logical systems and varieties of truth, therefore, certain universal patterns persist. Indeed, the aim of *controlling*, behind the back-and-forth of mathematical information, that information's *transference* (by way of functors, natural transformations, adjunctions, equivalences, etc.), constitutes one of categorical thought's main motivations. A subtle technical calculation over adjunctions yields to various *complex systems of gluing* between mathematical objects and allows us to better understand what Thom called the 'founding aporia of mathematics'.

From a pragmaticist point of view, we can amplify our conception of logic and open ourselves to a lattice of *partial flows of truth* over a *synechistic ground* (*syneches*: the glued, the continuous; *synechism*: the continuity operative in the realm of nature, and hypothesis of the Peircean architectonic). This indeed takes place with Yoneda's lemma **[247]**, where, in trying to capture a given reality (the category C, or, equivalently, its representable functors), the *forced emergence* of ideal objects (*non*representable functors) *necessarily widens and amplifies* the channels through which the functors at stake must flow. This ideal-real amplification is one of transitory ontology's strong points, consonant with the mixture of realism and idealism present in Peirce's philosophy. If we return to the pragmaticist maxim and its expression in figure 4 **[118]**, the diagram first places, on the left, a sign in an abstract category, and, on the right, the same sign partially incarnated in various concrete categories. The various 'modulations' and 'pragmatic differentials' let us pinpoint a sign that is one, abstract and general, and convert it into the multiple, the concrete and the particular. It is a task that, in category theory, is achieved by means of the various functors at stake, which, depending on the axiomatic richness of each of the categorical environments on the right, incarnate the general concepts in more or less rich mathematical objects.

This first process of specializing to the particular, concretizing the general, and differentiating the one, can thus be understood as an abstract *differential calculus*, in the most natural possible sense: to study a sign, one first introduces its *differential variations* in adequate contexts of interpretation. But, from the point of view of the pragmaticist maxim, and from the point of view of category theory, this is only the first step in a *pendular* and dialectical process. For once the variations of the sign/concept/object are known, the pragmaticist maxim urges us to *reintegrate* those various pieces of information into a whole that constitutes knowledge of the sign itself. Category theory, likewise, tends to show that behind the concrete knowledge of certain mathematical objects, and between these objects, exist strong functorial correlations (adjunctions, in particular), which are what actually and profoundly inform us about the concepts at stake. In either of these two approaches, we are urged to *complete* our forms of knowing, following the lines of an abstract *integral calculus*, the pendular counterpart of differentiation, which allows us to find certain approximations between various concrete particulars – particulars that may appear disparate, but which respond to natural proximities on an initially imperceptible, archeal ground.

The differential/integral back-and-forth, present in both the pragmaticist maxim and the theory of categories, is situated not only on the epistemological level mentioned above, but *continuously extends* to the 'what' and the 'where' of the (quasi-)objects at stake. The vertical interlacings to the right of figure 4 **[118]** – denoted by 'correlations, gluings, transferences', and situated under the *general sign* of the 'pragmatic integral' $(\int)$ – codify some of the most original contributions coming both from a broad modal pragmatism and from category theory, something we have already corroborated with sheaf theory. Both (mathematical) (quasi-)objects and (Peircean) signs live as *vibrant and evolving webs* in those environments of differentiation and integration. Sheaves, categories and pragmaticism therefore seem to answer to a complex regime of the prefix *TRANS*, on both the ontological and the epistemological level. We will now make our way into that second dimension.

CHAPTER 9

COMPARATIVE EPISTEMOLOGY AND SHEAFIFICATION

We have described the (quasi-)objects of contemporary mathematics as *webs* of structured information and as (in Leibniz's sense) 'compossible' *deformations* within those webs, open to *relative and asymptotic* composition in variable contexts. The dynamism of those webs and deformations straddles the eidal and the quiddital, in an iterated weaving of conceptual and material approaches. These are *bimodal* objects in Petitot's sense, situated in both physical and morphological-structural space, acting and reacting along spectra of formal and structural correlations with the many mobile environments where their partial *transit* takes place. In this way, the ontological 'what' and 'where' are blurred, and their frontiers become murky. We thus confront an *ontological fluctuation* that may provoke a predictable *horror vacui* in certain analytic approaches to the philosophy of mathematics, which seek to delimit and pinpoint their perspectives in the clearest possible way, fleeing from smears and ambiguities, and situating those tidy delimitations over fragments of the *absolute*. Nevertheless, for a contrapuntal, synthetic approach, open to *relative* transit, as heralded by contemporary mathematics, it becomes imperative to consider the *mobile frontiers* between the conceptual and the material.[349] A transitory ontology, as described in the previous chapter, thus gives rise to a fluctuating ontology, one that is not easily pigeonholed into the watertight compartments of Shapiro's square **[10]**: a natural *variation* of the 'what' and the 'where' gives rise to an associated variation of the 'how'. Once the correlative spectrum of disparate epistemological perspectives is opened up – in what we could call a *comparative epistemology* – we will, in this chapter, go on to reintegrate several of those perspectives in a sort of *epistemological sheaf*, sensitive to the inevitable complementary dialectic of variety and unity that contemporary mathematics demands.

In part 2, we repeatedly saw how, *behind* the differentiated, many of mathematics' most pivotal constructions bring *pendular* processes of differentiation and reintegration into play, beneath which emerge invariant, archeal structures. Recall, for example, Grothendieck's motifs beneath the variations of cohomologies **[144–148]**, Freyd's classifier topoi beneath the variations of relative categories **[245–246]**, Simpson's arithmetical

---

349 For Petitot, 'there is a rational solidarity between the conceptual, the mathematical, and the experiential, which challenges the positivist conception of the sciences and leads to a rehabilitation of critique on new bases' (protogeometries, morphological-structural order, local/global dialectics, phenomenological invariants of the world and not only of language, etc.). See Petitot's contributions to the *Enciclopedia Einaudi*, and in particular the entry, 'Locale/globale', *Enciclopedia Einaudi* (Torino: Einaudi, 1979), vol. 8, 429–90, and 'Unità delle matematiche', *Enciclopedia Einaudi* (Torino: Einaudi, 1982, vol. 15), 1,034–85 (p. 1,084 cited above). In the latter, Petitot places Lautman at the center of his argument ('La filosofia matematica di Albert Lautman', ibid, 1034–41): this was the first deep presentation of Lautman's work outside France.

nuclei beneath the theorematic variations of 'ordinary' mathematics [248–250], Zilber's proto-geometric nuclei beneath the variations of strongly minimal theories [255–257], the Lax-Phillips semigroup beneath the non-euclidean variations of the wave equations [218–220], the Langlands group beneath the variations of the theory of representations [184–185], the Grothendieck-Teichmüller group beneath arithmetical, combinatorial and cosmological variations [225, 233], Gromov's h-principle beneath the variations of partial differentials, etc. In all these examples, through webs and deformations, *the knowledge of mathematical processes advances by means of series of iterations in correlative triadic realms: differentiation-integration-invariance, eidos-quidditas-arkhê, abduction-induction-deduction, possibility-actuality-necessity, locality-globality-mediation.*

Advanced mathematics invokes this incessant 'triangulation', and perhaps finds its most (technically and conceptually) striking reflection in the notion of sheaf [285–288]. A fact of tremendous importance in contemporary mathematics is the *necessity of situating oneself in a full-fledged thirdness without reducing it* (without 'degenerating it', Peirce would say) to secondnesses or firstnesses. If thirdness disappears from the analytic modes of understanding, this is because existing (classical) forms of analysis are basically dual, as either-or exclusions, in the style of Benacerraf [11]. This means that the analytic viewing apparatus finds itself *intrinsically limited* when it comes to observe how advanced mathematics operates, *if it considers it on its own*, *independently of any synthetic contrast*. A pendular combination of the analytic and the synthetic, the differential and the integral, the ideal and the real, seems to be the epistemological path to follow. Let us note in passing that, on a 'meta-epistemological' level, what is needed is a *primacy of the synthetic*, not the analytic, in order to allow flows, oscillations and gluings between analytic and synthetic subfragments on lower levels.

The usual 'idealist' or 'realist' epistemologies, as they have been *independently* presented in the first part [6-9,124], are burdened with exclusions that do not suit contemporary mathematics' decantation of the ideal-real. *Neither* an analytic-ideal differentiation (which would, for example, block the emergence of Grothendieck's schemes [168–169]), *nor* a synthetic-real integration (which, for instance, would block the emergence of Gromov's inequality webs [260–261]), can cover the world independently. When it is a question of drawing together a collection of ever-finer[350] *recoverings*

---

350 Those recoverings can be seen as worthy, *metaphorical* analogues of Grothendieck topologies [138] and Voevodsky's algebraic surgery [146, 258]. On the role of metaphor, *both* in mathematical invention ('what') *and* in the subsequent knowledge of that invention ('how'), see the following chapter. In a certain sense, the real can *only* be covered by introducing metaphorical images, in the same manner that, in Yoneda's lemma [244–249], a category can be 'really' covered (with its associated category of presheaves) by introducing 'ideal' or 'imaginary', nonrepresentable, objects.

of the ideal/real junction, a 'comparative epistemology' becomes the order of the day. Articulation, dialectical balance, correlativity, and pragmatics here turn out to be indispensable. If every epistemological perspective generates an interpretive *cut*, a peculiar *environment of projectivity*, a differential modulation of knowing, then the next step consists in articulating the coherent projectivities, balancing the polarities, and gluing the modulations, *just as Peirce's pragmaticist maxim postulates* **[112–120]**. As we enter this process we become immersed in a sort of epistemological 'sheafification', where the local differential multiplicity is recomposed into an integral global unity.

Before considering the theoretical obstructions and advances that emerge in that enterprise of *sheafifying a comparative epistemology*, it may be useful to consider a detailed example. Consider the processes of 'smoothing' and 'globalization' that Gromov introduced into Riemannian geometry **[259–260]**. On the one hand, Gromov studies infinitesimal deformations of inequalities in well-defined *local* contexts; on the other, he studies the set of those deformations according to many possible metrics over the manifold, and calculates *global* invariants tied to a consideration of all the metrics together. In this back-and-forth, the knowledge of the objects is situated *neither* in the process of locally deforming and circumscribing the inequalities *nor* in the process of constructing global invariants, *but rather* in an indispensable dialectic between the two approaches: without the web of analytic inequalities, the synthetic invariants do not emerge [260–262]; and without the invariants the web of inequalities drifts about pointlessly (whether this be from a technical or a conceptual point of view). The knowledge of Riemannian geometry thus incorporates both analytic elements and synthetic configurations, situating itself in both 'idealist' (set of all metrics) and 'realist' (local physical deformations) perspectives *simultaneously*. In the case in question, the *sheaf* of the different local metrics gives rise to continuous sections tied to the invariants at stake. Going beyond this one case, however, the same Gromov points out [262] how the contrast between the exponentials, clouds, and wells in Hilbert's tree governs the *multidimensionality* of mathematical knowledge, which is *never reducible* to just one of its dimensions.

As we have seen in the generic analysis of the notion of sheaf [283–288], a sheafification in comparative epistemology requires us firstly to *prescind* the notions of neighborhood, covering, and coherence, before moving on to possible gluings. The notion of a *neighborhood between epistemological perspectives* requires us to postulate, first of all, the mutual *commensurability* of the latter; here, we explicitly position ourselves against the supposed incommensurabilities of 'paradigms' (Kuhn), a phenomenon that *rarely occurs*, at least in mathematics.[351] In this context, an *epistemological neighborhood* should bring together different perspectives, *either in their methods* (for example, the 'synthetic' method applied to a realist or idealist stance: pigeonholes 2_3 and 2_4 in figure 5 **[125]** regarding the 'pure' posing of problematics in the philosophy of mathematics , where i_j denotes the j_{th} pigeonhole in column i), *in their objectives* (in their approach to the real, for example: pigeonholes 1_3 and 2_3 in figure 5), *or in their mediations* through the warps at stake (for example, asymptotic mediation: pigeonholes 1_6 and 2_5 in figure 5). This *third option* is the most rich and innovative from a conceptual point of view, since it can help us eliminate dualistic exclusions by working with epistemic *webs of approximation*.

Once a certain neighborhood has been established (pursuing the cases above, a neighborhood of 'synthetic methodology', 'realist finality' or 'asymptotic mediation', for example), we pose the question of how that neighborhood may be *covered*. The neighborhoods could be covered in a binary fashion (by means of the pairwise-indicated pigeonholes, in the cases above, for example), but the most interesting cases might correspond to *non*binary

---

351 From an *epistemological* point of view, not only do the various paradigms ('isms') in mathematics (logicism, intuitionism, formalism, structuralism, etc.) not cancel each other out, in fact, they benefit from *mixed fusions* (through their reinterpretation in terms of category theory, for example). From a *logical* point of view, the *amalgams* in model theory are legion, and many interesting works in the area correspond to the crossing of *apparently* incommensurable entities.

coverings, as happens in the third situation, for instance: the 'asymptotic mediation' can be covered not only with a 'diagonal' binary (pigeonholes 1_6, 2_5 of figure 5), but with a ternary 'corner' (pigeonholes 1_6, 2_5, 2_6 of figure 5). The emergence of this 'corner', beyond a merely dual counterpoint, leads us to observe the *importance of the limiting and the asymptotic*, a crucial condition not only in the extrinsic knowledge of objects, but in the intrinsic characteristics of the investigated objects themselves (condition (ix) regarding the specificities of contemporary mathematics **[42]**).

In some cases, the coverings might be *coherent* (that is to say, not locally contradictory) and might thereby give rise to adequate *gluings* (*epistemological sections* in the sheaf). These gluings might open new epistemological perspectives, capable of locally responding to certain problematics in one way and other problematics in *another* way, so as to preserve the coherence of the responses. This would give impetus to a *partial and asymptotic use of relative epistemological strategies* (that change with respect to changes of neighborhood), *going beyond the restrictive or absolute epistemologies* of the sort proposed in Shapiro's square **[10]**. Let us consider a concrete case of this situation, by asking *how* we can know the structure of groups. Its archeal root, as a phenomenological invariant in the *world* (and not only as a descriptor in a *language*), surfaces through the works of Grothendieck **[144–146]**, Connes **[225–227]**,

Kontsevich [232–233], Gromov [264–265], Zilber [255–256], etc. Nevertheless, genetically, the notion of group appears as an eidal organization of particular symmetries (with Galois), which later (with Jordan) yields enormous quiddital facilitations in calculation. A group is thus an archetypical (quasi-)object, *at once* both ideal and real. The *apparent* confusion of these three perspectives is eliminated, however, by reading them as fragments of a section in a sheaf. If, on the one hand, we identify the topological base of a sheaf with a collection of *neighborhoods in a temporal tree* ('history'), then we see how a group – locally symmetrical in one neighborhood (Galois, 1830), locally combinatorial in another (Jordan, 1870), locally structural in another (Noether, 1930), locally cohomological in yet another (Grothendieck, 1960), or locally cosmological in still yet another neighborhood (Connes, Kontsevich, 2000) – can perfectly well live in multiple registers of knowledge, *each coherent with the others*. If, on the other hand, we take the base of the sheaf to be a collection of *neighborhoods in a conceptual map* ('geography'), then we see how a group – locally linear in one neighborhood (Galois, Jordan, Noether), locally differential in another (Lie, Borel, Connes), locally arithmetical in yet another (Dedekind, Artin, Langlands), or locally categorical in still yet another neighborhood (Grothendieck, Serre, Freyd) – can be *projected* onto all of those apparently divergent manifestations. 'History' and 'geography' can, in turn, be founded in a sort of *squared sheaf* that traces out all of the richness, both extrinsic and intrinsic, of the concept.

In accordance with the multidimensionality of mathematical vision, with the depth of 'Hilbert's tree', with relativity of the webs of categorical perspectives, and with those webs' sheafification, we can sense the existence of the *complex protogeometry* that underlies a comparative mathematical epistemology. Let us make explicit here, behind these considerations of ours, the hypothesis of a continuity[352] between the world of phenomena, the world of mathematical (quasi-)objects associated with those phenomena, and the world of the knowledge of those objects – which is to say, the *hypothesis of a continuity between the phenomenal, the ontic and the epistemic*. The mathematical constructions (and discoveries) that we have extensively reviewed in this work's second part show that contemporary mathematics offers *new support* for the possible soundness of this continuity hypothesis. Advanced mathematics, here, offers a finer illustration than elementary

---

352 This continuity is *one* expression of Peircean *synechism*, which postulates an even stronger continuity hypothesis, by supposing the existence of a *completely operative* continuity in nature (in which the human species appears, according to Peirce, both materially *and* semiotically). Another expression of this synechism is constituted by the three cenopythagorean, universal categories, which, according to the Peircean hypothesis, continually traverse both the world of phenomena *and* the forms by which those phenomena are known. For a description of synechism, of the generic (nonclassical, nonCantorian, nonextensional) concept of the continuum according to Peirce, and of certain partial mathematical models for this nonstandard continuum, see F. Zalamea, *El continuo peirceano* (Bogota: Universidad Nacional, 2001). An English translation is available at http://unal.academia.edu/FernandoZalamea/Websites.

mathematics, since the latter's low thresholds of complexity *impede* the emergence of the continuous/discrete dialectic that ceaselessly traverses the contemporary space of mathematics. From an epistemological point of view, the distinct perspectives are nothing other than *breaks in continuity*. In those breaks (as in Peircean abduction), *new* forms of knowledge are generated,[353] and – in an epistemology open to transit – those forms of knowledge, when they are coherent, can be subsequently reintegrated in an adequate fashion.

The *protogeometry* underlying a comparative epistemology of mathematics exhibits several peculiar characteristics, tied to a *combinatorics of coherent couplings* between the webs in play (deep, multidimensional, iterative webs). On the one hand, in fact, the *inverse bipolar tensions* between prescission and deduction **[286]** show that, in many mathematical cases past a high threshold of complexity (among which we may place the case of sheaves, in any of their geometrical, algebraic or logical expressions), there emerges a ('horizontal') hierarchy of partial couplings, whose *striated* resolution yields important forms of mathematical knowledge. This is the case, for example, with Gromov's *h*-principle **[263]**, where an 'inverse bipolar tension' between local and holonomic sections yields an

---

353 There may exist, here, a profound analogy between the processes of *symmetry breaking* in the physics of the first instants of the universe, and the processes of continuity breaking in the continuous archeal groups by which those forms of symmetry can be represented.

entire series of homotopic mediations, with calculative fragments of enormous practical interest in the resolution of partial differential equations. On the other hand, the *self-referential exponential nodes* in Hilbert's tree [262] yield *another hierarchy* (now 'vertical') of partial couplings whose *stratified* resolution likewise results in remarkable advances. This, for example, is what happens with the fundamental notion of space, whose breadth expands as it transits between sets, topological spaces, Grothendieck topoi [140], and elementary topoi [195], on each stratum generating *distinct* (though *mutually coherent*) conglomerates of new mathematical results. But there also exists at least a *third hierarchy* (now 'diagonal') of partial couplings between mathematical webs, directly influenced by the *transgressive* spirit of Grothendieck. Beyond the horizontal or vertical displacement, and *beyond their simple combination*, there do indeed exist diagonal mediations with archeal traits ('nondegenerate' Peircean thirdnesses), between remarkably far-flung realms of mathematical knowledge (dessins d'enfants between combinatorics and complex analysis [225], the Grothendieck-Teichmüller group between arithmetic and cosmology [225], noncommutative groups between logic and physics [258], etc.).

The protogeometry of those couplings thus incorporates an entire complex interconnection of multidimensional elements, in concordance with the 'intuitive' multidimensionality of mathematical knowledge.

The *image* of that mathematical knowledge is far removed from its logical foundation,[354] and a new preeminence of geometry appears on the map (item 7 in the distinctive tendencies of contemporary mathematics [41]). The beginning of the twenty-first century may indeed be a good time to begin to seriously consider a *geometricization of epistemology* such as we are proposing,[355] that would help us to overcome (or, at least, to complement) the *logicization of epistemology* undertaken throughout the twentieth century. The influence of *analytic philosophy*, whose logical support boils down to first order *classical logic*, should be countered by a *synthetic philosophy*, one much closer to the emerging *logic of sheaves*. What is at issue here is an important paradigm shift in mathematics *and* logic (as in Caicedo's fundamental results [283]), which philosophy (ontology, epistemology, phenomenology) *too* should begin to reflect. Indeed, the changes in

---

354 This is something that impresses itself upon even such a studious partisan of foundations as Feferman: 'The logical picture of mathematics bears little relation to the logical structure of mathematics as it works out in practice'. S. Feferman, 'For Philosophy of Mathematics: 5 Questions', 13. Course material, http://math.stanford.edu/~feferman/papers/philmathfive.pdf. Feferman nevertheless repeats the usual prejudices against an 'ingenuous' or 'trivial' Platonism: 'According to the Platonist philosophy, the objects of mathematics such as numbers, sets, functions and spaces are supposed to exist independently of human thoughts and constructions, and statements concerning these abstract entities are supposed to have a truth value independent of our ability to determine them' (ibid., 11). Compare this (caricatural) description with the more complex Platonism of a Lautman [52–60] or a Badiou [277–280].

355 Petitot's program for the *naturalization of phenomenology* covers similar bases, and offers a great deal of room to geometry. See J. Petitot et al., *Naturalizing Phenomenology: Issues in Contemporary Phenomenology and Cognitive Science* (Palo Alto: Stanford University Press, 2000). Although Petitot makes use of techniques in neuroscience, which we do not mention here, his invocation of Riemannian geometry and sheaf logic anticipates our own perspectives.

the logical base register the deformability of mathematical (quasi-)objects through relative transits, and allow us to overcome the classical 'rigidity' an object exhibits in a supposedly absolute universe.

In a detailed approach to Grassmann's work, Châtelet lucidly describes one of the fundamental problematics of mathematical knowledge:

> Like the contrast continuous/discrete, the equal and the diverse are the result of a polarization; it is thus that algebraic forms 'becoming through the equal' and combinatory forms 'becoming through the diverse' can be distinguished. *It is a matter of finding the articulation that makes it possible to pass continuously from the equal to the diverse.*[356]

Châtelet's mathematico-philosophico-metaphorical instruments in this search for a *continuous* articulation include 'dialectical balances', 'diagrammatic cuts', 'screwdrivers', 'torsions', and 'articulating incisions of the successive and the lateral', which is to say, an entire series of *gestures* attentive to movement and which 'inaugurate dynasties of problems'[357] and correspond to a certain *fluid*

---

356 G. Châtelet, *Les enjeux du mobile*, 167 (our emphasis).

357 Ibid., 37, 33, 38, 218, 32.

electrodynamics[358] of knowing. Here, we are once again facing the multidimensionality of mathematical (quasi-) objects, a multiplicity that can only be apprehended by way of 'gestures', that is, by way of *articulations in motion*, which allow for partial overlaps between the 'what' and the 'how'. Mathematical epistemology – in so far as it wishes to be able to incorporate the objects of contemporary mathematics (and in many cases, those of modern mathematics) into its spectrum – should therefore be essentially mobile, liable to torsion, capable of reintegrating cuts and discontinuities, and sensitive to fleeting articulations – in sum, *genuinely in tune* with the objects it aspires to catch sight of. No fixed position, determined a priori, will be sufficient for understanding the *trans-form*-ability of the mathematical world, with its elastic transits, its unstoppable weavings between diverse forms, and its zigzagging pathways between modal realms.

In the case studies in part 2, we can concretely detect various properties of the *epistemic protogeometry* that we have just been discussing. *Both* in the (quasi-)objects at stake, *and* in the forms by which they are known, we observe common protogeometrical features, among which we must emphasize: (*a*) multidimensional cuts

---

358 Châtelet interlaces Grassmann's 'fluidity' (ibid., 166) with Maxwell's electromagnetism in order to study an *electrophilosophy* in proximity with Faraday's electrogeometrical space (ibid., chapter 5). A geometrical perspective on Maxwell and Faraday's '*allusive* operators' (ibid., 219; emphasis ours) breaks with punctual or instantaneous interpretations, and explores the asymptotic deformations of entities within given neighborhoods ('the pedagogy of lines of force', ibid., 238–48).

and reintegrations, (*b*) triadic iterations, (*c*) deformations of objects and perspectives, (*d*) processes of passage to the limit by means of nonclassical approximations, (*e*) asymptotic interlacings and couplings, and (*f*) fragments of sheafification. In contemporary mathematics, there is a *continuous overlapping* of the 'what' and the 'how', whose very protogeometry tends to emerge from a common and *unitary* base. Grothendieck's schemes **[139–140]** combine many of the above properties: 'ontically', the schemes are constructed by (*a*) introducing structural representations of rings, (*c*) viewing points as prime ideals and topologizing the spectrum, (*f*) locally defining fibers as structural spaces of rings and gluing them together in an adequate fashion; 'epistemically', schemes involve (*a*) a process of generalization and specification between multidimensional objects, (*b*) an iterated threeness[359] between a base, its widening, and their mediation (projection), (*c*) an comprehension of objects through their relative positions, etc. Similar registers can also be specified for Grothendieck's other great constructions: topoi **[141–144]** and motifs **[144–148]**. In the same way, the Langlands correspondence **[181–186]** includes, 'ontically', (*a*) representations of groups, (*b*) iterated

---

359 The *first level* of iteration corresponds to the (algebraic) step from the initial, unitary, commutative ring to the set of its prime ideals; the *second level*, to the (topological) step from the combinatorial spectrum to the spectrum with the Zariski topology; the *third level*, to the (categorical) step from neighborhoods to fibers; the *fourth level*, to the ultimate sheafification of the whole. It is interesting to observe how those iterations *are not absolute*, and can be realized in a different, or better still, *mixed* order.

transits between the modular, the automorphic and the *L*-representable, (*c*) mixed differential and arithmetical structures, and (*d*) manipulations of noncommutative groups; 'epistemically', the strategy of the program rests on (*a*) an understanding of arithmetical objects as 'cuts' (projections) of more complex geometrical objects, (*b*) a systematic search for geometrical mediations between (discrete) arithmetic and (continuous) complex analysis, and (*c*)-(*d*) an observation of the objects' analytical deformations, etc. Another similar situation occurs with the general theory of structure/nonstructure according to Shelah **[196–197]**, where, 'ontically', there arise (*a*) amalgams in high finite dimensions, (*d*) cardinal accumulations by means of nonexponential objects (PCF theory), and (*e*) monstrous models and asymptotic warps of submodels; 'epistemically', Shelah's vision integrates (*a*) a celebration of mathematics' multidimensionality, (*b*) an incessant *moderation* between the structured and the nonstructured, (*d*)-(*e*) a deep understanding of the objects at high levels of the set-theoretic hierarchy as limits of 'moderate' and 'wild' fragments, etc. In this way, certain common protogeometric characteristics *naturally* lace together the ontic webs and the epistemic processes in play.[360] Throughout part 2 of this work, we have *implicitly*

---

360 In the last instance, the transformations leading from a fixed ontology to a *transitory ontology*, and from a fixed epistemology to a *sheafified, comparative epistemology*, cause the 'entities' under investigation in each of these approaches ('what', 'where', 'how') to draw closer to one another, and cause their mobile frontiers to become much less exclusive.

carried out other analyses, and additional examples could be explicated here (the works of Connes, Kontsevich, Zilber, and Gromov, especially, lend themselves to this), but the aforementioned cases may provide sufficient illustration.

The principal limitation that seems to encumber any 'analytic' epistemology, as opposed to the 'synthetic', comparative epistemology that we are hinting at here, pivots on the analytic difficulty in confronting certain inherently vague environments, certain *penumbral zones*, certain 'outposts of the obscure', as Châtelet calls them,[361] certain elastic places of 'spatial negativity', certain 'hinge-horizons' where complex mixtures emerge that *resist* every sort of strict decomposition. We will study the problematic of the penumbra with greater care in the next chapter, where we take up the (sinuous, nonlinear) dynamics of mathematical *creativity*, but in what remains of this chapter we will approach the dialectic of the obscure and the luminous in conjunction with the three ubiquitous polarities, *analysis/*

The ontic 'webs' and epistemic 'processes' are therefore only relative specifications (in ontological and epistemic contexts) of a single and common kind of '*proto-actions*' (something which coincides with certain tendencies of Peircean universal semiotics). In such a 'blurring' of frontiers between the ontic and the epistemic, it is worth pointing out how Badiou, on the one hand, sees mathematics as being, basically, ontology, while Petitot, on the other hand, considers it to be, basically, epistemology. On an analytic reading, such blurrings are taken to be improper, but, as we have observed, from a synthetic reading – *once transits, osmoses, and contaminations are accepted* – new *analyses* of the processes of transference can be carried out, without having to spill over into an extreme relativism or into ingenuous forms of skepticism. The forms of the decomposition (analysis) of transit (synthesis) can no longer be forgotten in mathematical philosophy.

361 Châtelet, *Les enjeux du mobile*, 22, 37.

synthesis, *idealism/realism* and *intensionality/extensionality*, which appear in any epistemological approach. Our objective consists in *mediating* (or 'moderating': Grothendieck, Shelah) between them, and proposing reasonable couplings from the 'outposts' of contemporary mathematics.

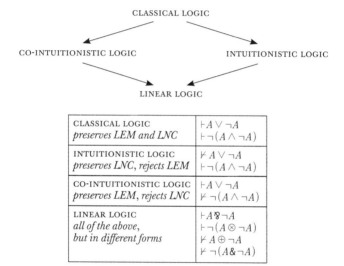

Figure 14. *Logic: classical, intuitionistic, co-intuitionistic, and linear.*

A very interesting case of this search for *mediations in penumbral zones* corresponds to the dualization of the central *Clas − Int* logical axis (the continuum of superintuitionistic logics between intuitionistic logic and classical logic) into its counterpart *CoInt − Linear* (logics between linear logic and paraconsistent, co-intuitionistic logic). The *Class − Int* axis can be seen as a line along which the law of the excluded middle progressively gives way, by continuous degrees. Now, the law of the excluded middle has a *dual*, which is the *principle of noncontradiction*. In the eyes of classical logic, these two laws are equivalent, but the *Class − Int* axis progressively breaks this symmetry, dissolving the excluded middle while maintaining non-contradiction. What this suggests, then, is the possibility of a *dual* axis, that would evaporate noncontradiction while maintaining the excluded middle, terminating a logical system that is the dual of intuitionistic logic: a 'paraconsistent' system called *co-intuitionistic logic*.

What these two logics sacrifice in symmetry, they gain in unearthing a coherent and profound *computational dynamism*, allowing for a mapping of proofs onto functional *programs*, and the dynamic of *cut-elimination* (the procedure by which their proofs can be reduced to a unique normal form, a 'direct proof' where no *lemmas* are used) onto the process of program execution. Now, as the French proof-theorist Jean-Yves Girard has observed, the transformation of classical logic into intuitionistic logic can be seen as the result of an asymmetrical restriction of the 'structural rules' that govern the former – rules which allow formulas to be reused and forgotten as needed in the course of a proof, and so encode their measure of *ideality*. A similar observation can be made for the shift from classical to co-intuitionistic logic. As Girard observes, it is therefore possible to recover *both* the coherent dynamics of intuitionistic (and co-intuitionistic) logic, *and* the protogeometrical symmetry of classical logic, by *symmetrically suspending the structural rules* – which gives us *linear logic*, a logic where formulas work more like transient *actions* (and their negations as *reactions*) than like atemporal and ideal *propositions*. The forms of conjunction and disjunction, here, are each split in two – into 'multiplicative' and 'additive' forms – which allows this *synthetic* logic to partially preserve and partially reject *both* the law of the excluded middle *and* its dual, the law of noncontradiction. The power of the structural rules, alone capable of reuniting these forms, is then reintroduced in the form of *local modal operators*, which, in fact, constitute some of the most stubborn *obstructions* to geometrical and dynamical analyses of linear proofs (Girard calls them the 'opaque modal kernels' of 'essentialism').[362]

The diamond of the four logics[363] complete the primordial *Class – Int* axis in a natural manner, and Luke Fraser has posed the fundamental question as to what concepts would correspond to the well-known landscape

---

362 J.-Y. Girard, *The Blind Spot: Lectures on Logic* (Zurich: European Mathematical Society, 2011), 11.

363 The 'diamond' is outlined and briefly discussed in Samuel Tronçon's outstanding genealogy of the Girardian research program in S. Tronçon. *Dynamique des démonstrations et théorie de l'interaction*, PhD thesis, University of Aix-Marseille I, Marseille, April 2006, 278.

Class (analysis, discreteness, extensionality, sets) – *Int* (synthesis, continuity, intensionality, categories) if we attempt to *dualize them toward the CoInt – Linear side*. Indeed, the *penumbral zone* is all the more accentuated given that, in mathematics' current state of development, we do not seem to be able to even glimpse natural completions of the dyads of analysis/synthesis, discrete/continuous, extensionality/intensionality, and sets/categories. If we think of Thom's 'founding aporia' as describing the conceptual galaxy that turns around just *one* of these axes (*Class – Int*), this leaves, in the penumbra, three axes around which may revolve *entire galaxies*, conceptual spaces of which we have not yet learned to dream.

Lawvere has introduced limit co-operators and co-Heyting algebras in topoi, the Brazilian school has worked on categorical models for paraconsistent (co-intuitionistic) logics and, recently, the Dutch school has given impetus to the study of co-algebraic techniques in computational modelings. The analyses carried out, nevertheless, have turned out to be very localized and correspond to the resolution of minor technical questions. A *fundamental philosophical* study of the galaxies associated with the penumbral zone *CoInt – Linear*, by contrast, could, in the coming years, produce a genuine *conceptual revolution* – yet another example of how contemporary mathematics can serve to tune the strings of philosophy and make them vibrate. The triadic Peircean phenomenology could be of great assistance here, at least in relaxing the dyads so as to yield a third opening (with Giovanni Maddalena we have, for example, proposed an extension of *analysis/synthesis* into *analysis/synthesis/horosis*,[364] through the concept of *horos*, border, frontier), but the labor of conceptual completion through the 'Fraser Side', $CoInt - Linear$ still entirely remains be to done.[365]

The analysis/synthesis polarity has always been a source of equivocations and advances, in philosophy[366] no less than in mathematics.[367] In a dialectical fashion (of the thesis-antithesis type, $\theta\,\bar{\theta}$) the polarity brings into opposition knowledge by decomposition and knowledge by composition, but *in a relative fashion*, since 'there exist no absolute criteria, either for processes of analysis or for processes of synthesis'.[368] Indeed, nature always presents itself by way of *mixtures* that the understanding decomposes

---

364 G. Maddalena & F. Zalamea, 'A New Analytic/Synthetic/Horotic Paradigm. From Mathematical Gesture to Synthetic/Horotic Reasoning,' (*preprint*, submitted to Kurt Gödel Research Prize Fellowship 2010).

365 We are grateful to our translator, Zachary Luke Fraser, for having presented this problematic to us, and for helping us to develop our discussion of the 'logical diamond'.

366 See G. Holton, 'Analisi/sintesi', *Enciclopedia Einaudi* (Torino: Einaudi, 1977), vol. 1, 491–522. On many occasions we have tried to (technically, conceptually and analogically) *use* the *Enciclopedia Einaudi*, whose eminently *transductive* project quite naturally brings together modern and contemporary mathematics. The last two volumes (vol. 15, *Sistematica* and vol. 16, *Indici*) are of great service, as well as the fascinating *maps*, tables, and reading diagrams developed by Renato Betti and his collaborators, where the many osmoses and shiftings of contemporary thought are made manifest. (Recall Merleau-Ponty's philosophy of the 'shifting' **[277]**.) Petitot's remarkable contributions to the *Enciclopedia* introduced us to the local/global, centered/decentered, and One/Many problematics that, as we have seen, reemerge in contemporary mathematics.

367 See the anthology edited by M. Otte and M. Panza, *Analysis and Synthesis in Mathematics, History and Philosophy* (Dordrecht: Kluwer, 1997).

368 Holton, 'Analisi/sintesi', 502.

and recomposes in iterated oscillations. Analytic differentiation (typical in the mathematical theory of sets: objects known by their 'elements') can lead to a better resolution of certain problematics, though subdivision in itself (Descartes) does not assure us of any result (Leibniz).[369] Synthetic integration (typical of the mathematical theory of categories: objects known by their relations with the environment), for its part, helps to recompose a unity that is closer to the mixtures of nature, but, as with its counterpart, inevitable obstructions arise in its use. One should therefore try to bring about a '*continuous effort of balancing analysis and synthesis*' (Bohr),[370] something that meets with peculiar success in Peircean pragmaticism, and particularly with its pragmaticist maxim **[111–121]**. An *epistemological orientation* can then emerge, leading to an abstract differential/integral (analytic/synthetic) 'calculus', situated over a *relative fabric of contradistinctions, obstructions, residues and gluings* **[118]**. The case studies that we have carried out in contemporary mathematics show that such an orientation is never absolute, but only *asymptotic*, that it depends on multiple relativizations – eidal ascents and quiddital descents, 'first, reason climbs up with analysis, and then down with synthesis'[371]

---

369 Ibid., 503.

370 Ibid., 505.

371 Ibid. The oscillations of thought have been studied with *geometrical* instruments at least since Ramon Llull, *Libro del ascenso y decenso del entendimiento* (1304) (Madrid: Orbis, 1985).

– but also requires us to have enough points of reference (archeal invariants) to calibrate the relative movements.

Lying still further beyond the pendular analysis/synthesis equilibrium *invoked* by Bohr, contemporary mathematics hints at how that equilibrium can be *produced*. We have seen how the notion of *sheaf*, in a very subtle manner, combines the analytic and the synthetic, the local and the global, the discrete and the continuous, the differential and the integral **[285–288]**. In this way, the *sheafification of the analysis/synthesis polarity* generates a new *web* of epistemological perspectives, following the directions of contemporary mathematics. If, in fact, we look at the *diagrammatization* of the general concept of sheaf that appears in figure 13 **[286]**, we can see how, behind the double opposition, analysis/synthesis (the vertical arrangement in the diagram) and local/global (horizontal arrangement), lies a very interesting *diagonal mediation* that is rarely made manifest. Taking the *pivotal* analytic concept of 'covering' and modulating it by means of a synthetic 'section-preservation-projection-restriction' hierarchy (figure 13), the natural notion of a *transversal transform of the covering* emerges, whereby the open and closed[372] *strata* of the various representations (or 'covers')

---

372 The notion of 'covering' comes from the Latin *cooperire* (*operire* [close]; *cooperire* [close or cover completely], eleventh century). In counterpoint with *aperire* (open), the notion of covering thus includes, by way of its etymological roots, a conception of the *transit* between open and closed environments (a transit reflected in other derived frontiers: *operculum* [overlay]; *aperitivus* [aperture]).

involved are superposed on one another. The transform
– tied to a sort of protogeometry of position ('pretopos')
– *combines* the analytic capacity to cover through decompositions (whether with elements, neighborhoods, or
asymptotic approximations), and the synthetic capacity
to recombine the fragmented through variable contexts
(whether with partial couplings or more stable structural
gluings). We will call this type of sweeping mediator
between the analytic and the synthetic a *Grothendieck
transform*,[373] a sweeping reticulation particularly attentive
to the *relative mixes* between the concepts in play.

By its very definition, the Grothendieck transform
incorporates peculiar modes of knowing. Through the
transverse, it introduces a reticular warp over what contradistinctions, couplings and asymptotes stabilize. Through
the *covering*, it introduces a fluid dynamics (Merleau-Ponty's 'shifting [*glissement*]'), incarnated in the gerund itself:
the 'to be covering' over a situation ('geography') and
a duration ('history'). The Grothendieck transform thus
changes our epistemological perspectives by the simple
fact of *smoothing them* as a whole (Gromov [**259, 263**]):
integrating them into an *evolving tissue*, it desingularizes

---

373 Looking back at chapter 4, it seems clear that this transformational process of
mathematical concepts has always been present in the global, conceptual inventiveness
of Grothendieck. From a local and technical point of view, moreover, Grothendieck
topologies [**138**] help to incarnate, in a restricted manner, what we refer to here
as a 'Grothendieck transform'. Indeed, for *any* arbitrary site (any category with a
Grothendieck topology), the category of presheaves over that site gives rise to a category
of sheaves, through a general process of sheafification that corresponds, precisely, to
bringing about the mediations contemplated in the 'Grothendieck transform'.

punctual perspectives, in favor of a comparability that emphasizes the regularities, mediations and mixtures between them. This is evinced in the *mathematical practice* that we have reviewed in the second part of this essay. The transversality and 'wonderful mélange' found in Serre's methods [177], the transversal contamination in Langlands's program [184], the omnipresent dialectic of adjunctions and the evolution of objects in Lawvere [193], the covering moderation in Shelah's PCF theory [201–202], Atiyah's Index Theorem with its medley of eidal transversality (equilibrium between transits and obstructions) and quiddital coverage (from algebraic geometry to physics) [208–209], the harmonic analysis (precise technical form of covering transversality) applied to the non-euclidean wave equation in Lax [219], the noncommutativity that runs transversally over quantum mechanics in Connes [223], the quantizations whose transversal recoverings allow us to reconstruct classical structures in Kontsevich [232], the intermediate categories between regular categories and topoi, produced as transversal cuts (*Map, Split, Cor*) over free allegories in Freyd [246], the logical nuclei covering second order arithmetic in Simpson [250], the protogeometrical kernels covering strongly minimal theories in Zilber [255], and the transversality of the *h*-principle in Gromov [263] are all very subtle and concrete examples behind which lurk the mediations and smoothings of Grothendieck's

'transversal covering transform'. In all of these constructions, the epistemological vision of mathematics is one
that is flexible enough to *cover a hierarchical stratification of
ideality and reality*, dynamically *interchanging* the contexts
of interpretation or adequation of the (quasi-)objects
whenever the technical environment or creative impulse
demands.

These considerations help us to better understand
how – in contemporary mathematics and, more broadly,
in advanced mathematics [27] – the poles of idealism
and realism should not be considered as separate (discrete
exclusion, through an analytic approach), but in *full* interrelation (continuous conjunction, through a synthetic
approach). Indeed, a good understanding of the ideal/
real dialectic in advanced mathematics can be achieved
through the notion of an epistemological back-and-
forth.[374] The back-and-forth postulates not only a pendular
weaving between strata of ideality and reality, but, above
all, a *coherent covering* of various partial approximations.

---

374 The back-and-forth emerges in Cantor's proof concerning the isomorphism between
two dense, countable linear orders without limit points. The Cantorian technique
(formalized in a modern fashion by Hausdorff in 1914) uses *approximations* to the
isomorphism by way of a collection of partial homomorphisms that cover the sets
little by little (surjectively), while continuing to preserve the orders, and whose well-
behaved limit furnishes the desired isomorphism. Observe, here, the gerundial, the
asymptotic and the limit, which anticipate several of the themes taken up in this chapter.
The back-and-forth was later used by Fraïssé (1954) to characterize the elementary
equivalence between abstract structures (with arbitrary relations, beyond orders),
and by Lindström (1969) in his surprising characterization theorems for classical,
first-order logic (maximal with respect to the compactness and Löwenheim-Skolem
properties). Ever since, the back-and-forth has become an indispensable technique in
abstract model theory. See J. Barwise, S. Feferman, eds., *Model Theoretic Logics* [New
York: Springer, 1985].

A *direct* example of this situation is found in Lindström-style back-and-forth technique used in abstract model theory: the semantic (stratum of reality, constituted by a collection of models with certain structural properties) is understood through a series of syntactic invariants (strata of ideality, constituted by languages with other partial *reflection* properties), and more specifically, the elementary (real) equivalence is reconstructed through (ideal) combinatory coherences in a collection of partial homomorphisms that are articulated in the back-and-forth. Other *indirect* examples appear in the case studies that we have carried out: the progressive back-and-forth in the elucidation of the Teichmüller space's functorial properties according to Grothendieck [170], the structural amalgams by strata according to Shelah [196], the approximations of hyperbolic groups and the polynomial growth of groups according to Gromov [259], etc.

In these processes, mathematical (quasi-)objects' modes of creation, modes of existence, and the modes by which they are known are interlaced and *reflected* in one another (*general transitoriness between phenomenology, ontology and epistemology*). The relative (partial, hierarchized, distributed) knowledge of those transits therefore becomes an indispensable task for mathematical epistemology. *Beyond* trying to define the ideal or the real in an *absolute* manner (a definition that, from our perspective, reflects a *poorly posed problem*), the crucial task of mathematical epistemology should instead consist in describing, pinpointing, hierarchizing, decomposing and recomposing the diverse forms of transit between the many strata of ideality and reality of mathematical (quasi-)objects. Through the same *forces* that impel the *internal* development of the mathematical world, we have, for example, explicated some contemporary forms of transit with great expressive and *cognitive* power: protogeometrization, nonclassical approximation, sheafification, Grothendieck transformation, back-and-forth modulation.

Another important epistemological *opening* is found in *mediating* the usual 'extensional *versus* intensional' dichotomy, which *grosso modo* corresponds to the 'sets *versus* categories' dichotomy. One of the *credos* of Cantorian set-theoretic mathematics – and, for that matter, of traditional analytic philosophy – has been the *symmetry* of Frege's abstraction principle, introduced *locally* by Zermelo with his axiom of separation: given a set A and a formula $\varphi(x)$, there exists a subset $B = \{a \in A : \varphi(a)\}$, and so an equivalence obtains (locally, within the restricted universe A) between $\varphi(a)$ (intensionality) and $a \in B$ (extensionality). But, beyond the *credo* and its indisputable technical convenience, there is no philosophical or mathematical reason to prevent an *asymmetrization* of Frege's principle of abstraction.[375] A *non-Cantorian continuum*, for

---

375 Various mathematical texts consider that a *breaking of the symmetry* between extension and intension may turn out to be beneficial. From our perspective, this symmetry

example, seems to involve profound intensional characteristics that are impossible to achieve with an extensional modeling.[376] In the same manner, many of the intensional characterizations of mathematical objects and processes obtained in category theory offer new perspectives ('relative universals') that *do not coincide*[377] with extensional set-theoretical descriptions.

Though the extensional/analytic/set-theoretic influence has, until now, been preponderant in mathematics, its intensional/synthetic/categorical counterpart every day becomes more relevant, and a new '*synthesis of the analysis/synthesis duality*' is the order of the day. Of course, the *self-reference* just mentioned in quotation marks is not

---

breaking could correspond to a *deformation* of the local symmetries codified in Zermelo's separation axiom (symmetries that obtain fiber by fiber, but that should collapse as a consequence of slight deformations of those fibers). See, for example, J. Bénabou, 'Rapports entre le fini et le continu', in J. M. Salanskis, H. Sinaceur, eds., *Le labyrinth de continu* (Paris: Springer-Verlag, 1992), 178–89; E. Nelson, 'Mathematical Mythologies', ibid., 155–67; R. Thom, 'L'anteriorité ontologique du continu sur le discret', ibid., 137–43.

376 From the point of view of the axiomatic bases required to capture a generic continuum such as Peirce's (non-Cantorian) continuum, Zermelo's local separation axiom is an excessively demanding postulate. By contrast, giving precedence to the intensional offers, from the outset, an important support for the *inextensibility* of the Peircean continuum, which is to say, the impossibility of defining it through an accumulation of points. In effect, when we *asymmetrize* the axiom of separation, only certain formulas can give rise to classes, and the a priori 'existence' of points can be eliminated: singleton sets $\{x\}$ do not always exist, and only in certain, specific (constructible) cases can they come to be actualized. At the same time, by permitting the manipulation of contradictory intensional domains (in the potential) without having to confront the associated contradictory extensional classes (in the actual) that would trivialize the system, we achieve a greater flexibility in our generic approach (freed from the tethers to the actual) to the continuum. See Zalamea, *El continuo peirceano*, 84–6.

377 What we are dealing with here is a crucial, mathematical noncoincidence, although *logically* the terms may be equivalent. The mathematical richness of category theory, as we have seen, *does not reduce* to a logical counterpart in the fashion of 'topos theory' ≡ 'restricted set theory', but it is directed, rather, toward the discovery of symmetries and synthetic equilibria, *unobservable* in light of analytic decompositions.

only noncontradictory, but exerts a *multiplicative* force in a *hierarchy* of knowledge. Throughout chapters 8 and 9, we have tried to forge some mediating perspectives within the '*synthesis of the analysis/synthesis duality*'. By expanding our spectrum to more general cultural horizons, we will, in the two final chapters, see how the incessant mediations of contemporary mathematics are grafted, on the one hand, to the (local) 'creative spirit' of the mathematician, and, on the other, to the complex and oscillating (global) 'spirit of the age' in which we find ourselves immersed.

CHAPTER 10

PHENOMENOLOGY OF
MATHEMATICAL CREATIVITY

Traditional philosophy of mathematics tends to neglect the ways in which mathematical thought emerges. Several texts dealing explicitly with mathematical invention have come to us from such practitioners of the discipline as Poincaré, Hadamard, Grothendieck and Rota, but, curiously, the professional philosopher neglects the *phenomenology of mathematical creativity* as something foreign to his reflection. Nevertheless, in science, and, more generally, in every area of knowledge, the way in which knowledge emerges is (at least) as important as the knowledge itself. As Valéry reminds us, 'the interest of science lies in the *art* of doing science':[378] the art of invention and the practices associated with creativity constitute the true interest of science. This is all the more obvious in the realm of mathematics, the specificity of which is rooted in the incessant transit (*ars*) of concepts, proofs and examples between the possible (abduction), the necessary (deduction) and the actual (induction). Valéry, a true connoisseur of mathematics and an extraordinary investigator of creative modulations in the twenty-seven

---

[378] J. Prévost, P. Valéry, *Marginalia*, *Rhumbs et autres* (Paris: Editions Léo Scheer, 2006), 229 (Valéry's emphasis).

thousand pages of his *Cahiers*,[379] used to point out, for his part, that 'the origin of reason, or of its notion, is, perhaps, the *transaction*. One must transact with logic, on one side, with impulse, on the other, and, on yet another, with facts'.[380] The phenomenology of mathematical creativity should confront those transactions, those contaminations, those impurities, which are what ultimately afford us the entire richness of mathematics. The reduction of the philosophy of mathematics to the philosophy of 'logic' (according to the usual tendencies of analytic philosophy, as we saw when we opened the *Oxford Handbook of Philosophy of Mathematics and Logic* **[101–107]**, from which 'Mathematics' had vanished), or the very reduction of the philosophy of mathematics to questions of 'logic' and of 'facts' (according to the slightly broader tendencies that take stock of mathematics' interlacing with physics), are approaches that leave behind the indispensable creative 'impulse' to which Valéry refers. In this chapter we will try to elucidate that (apparently vague and undefinable) *impulse*, which responds, nevertheless, to a complex phenomenological web of catalysts and graftings of inventiveness that can be made explicit.

---

379 Facsimile edition: P. Valéry, *Cahiers* (Paris: Editions du CNRS) (29 vols.), 1957–61. Critical edition: P. Valéry, *Cahiers 1894–1914* (Paris: Gallimard) (9 vols. at present), 1987–2003. Thematic anthology: P. Valery, *Cahiers* (Paris: Gallimard/Pléiade) (2 vols.), 1973–4.

380 Prévost & Valéry, *Marginalia...*, 225 (Valéry's emphasis).

The nondualist instruments that afford us an adequate perception of creative transit in mathematics, at bottom, have been under our nose all the while, even if often poorly interpreted: namely, the work of Plato, understood as a study of the mobility of concepts, as a metamorphosis of knowledge, as a description of connections and interlacings, as a subtle analysis of the *inter-* and the *trans-* has always been there, under our nose, though it has often been poorly interpreted. Beyond certain trivial readings, there nevertheless stands the *dynamic* Plato of Natorp,[381] thanks to whom it has become impossible to not see 'the genuine, dynamic sense of the idea' that renders 'untenable the interpretation of ideas as things'.[382] The processual, nonstatic Plato, a Plato not fixed to a reification of the idea, a Plato whom Natorp recuperated at the beginning of the twentieth century, and to whom Lautman [56] and Badiou [279] would later return, seems to constitute the nondual, *mobile base* that mathematics requires: an apparent contradiction in terms – for the approaches customary to analytic philosophy of mathematics, the 'base' should not turn out to be mobile. Nevertheless, in the rereading of Plato that Natorp proposes, we see how 'the *logoi* do not have to be governed by

---

381 J. Servois, *Paul Natorp et la théorie platonicienne des Idées* (Villeneuve d'Ascq: Presses Universitaires du Septentrion, 2004). An excellent and brief introduction to Plato by Natorp can be found in F. Brentano & P. Natorp, *Platón y Aristóteles* (Buenos Aires: Quadrata, 2004).

382 Ibid., 120, 92 (in the order of the texts cited).

the *onta*', how 'the definitions of mathematics, in reality, define methods; in no way do they define existent things or simple properties inherent in such things', how '*kinesis*, the movement, the transformation, or the gait, let us say, of concepts' governs all of mathematics, how the Platonic theory of ideas refers to the 'method of its positions', to its 'becoming, mutation or peregrination', and how the thought of Plato 'installs itself in the relative position' and opens itself to the study of the 'correlative [...], of change [...] and transition'.[383] Mathematics – bound up with the study of *logoi*, methods, partial representations, and relative positions, as we have abundantly corroborated in the contemporary realm, a realm that oscillates between the eidal, the quiddital and the archeal – can thus assume at the outset, as its mobile base, this Platonic thought that is so alert to the transformability of concepts/proofs/examples. The *mobility of the base*, indispensable for understanding Grothendieck's work **[142]**, underlies Platonic philosophy from the beginning.

Later in this chapter, we will draw from the works of Merleau-Ponty, Blumenberg and Rota to further refine the 'mobile base' just indicated. But already – simply by way of *the possibility and plausibility* of the hierarchy of modulations and 'transactions' afforded by a dynamic Platonic philosophy – we can better understand the modes

---

383 Ibid., 30, 35, 36, 55, 57, 72–3.

by which mathematical creativity emerges. Indeed, we can already begin to situate the *tension between discovery and invention* under 'reasonable' presuppositions. Taking the word 'reasonability [*razonabilidad*]' in the sense given to it by Vaz Ferreira – as a gluing together of 'reasonable' and 'sensibility'[384] – we can say that some *reasonible* Platonic presuppositions underlying the invention/discovery polarity are: that the polarity is not antagonistic or dual, but, rather, *entwined in a web*; that through this web transit various types of quasi-objects, modalities, and *images*; that the *positions* of those quasi-objects, modalities, and images are not absolute, but relative; that a progressive *gradation* determines, depending on the context in question, the mathematical quasi-objects' proximity to each of the polar extremes; that in that gradation, observations of *structuration* tend to approach (in spiraling or asymptotic turns) the processes of discovery, while constructions of *language* tend to approach (after perhaps one more 'turn of the screw') processes of invention.

In the emergence of mathematical thought, *contaminations* are legion. Recall Grothendieck's magnificent text on motifs **[145–146]**. In it, mathematical creativity is distributed across a great variety of registers: the initial 'listening' to the motif (Peircean firstness), its incarnation

---

384 C. Vaz Ferreira, *Lógica viva* (Caracas: Biblioteca Ayacucho, 1979). On Vaz's 'reasonability', see A. Ardao, *Lógica de la razón y lógica de la inteligencia* (Montevideo: Marcha, 2000). The works of Vaz Ferreira (Uruguay, 1872-1958) open up very interesting (and untapped) natural osmoses between 'pure' and 'human' sciences.

in a 'multitude of cohomological invariants' (secondness), and its pragmatic interlacing through modulations of the 'basic motif' (thirdness). But the process does not stop there; it is not static, cannot be isolated, and *iterates itself recursively*: given a cohomology (first), we study the topological spaces (second) captured by *that* cohomology, and then determine the transits (third) between the spaces that the given cohomology codifies. And so, in succession: a position *relative* to a given space, a position *relative* to a given transit, and so on. Mathematics thus proceeds by means of maximal connections of information ('saturations', as Lautman would say) in *evolving strata* of knowledge. Creativity emerges through that variable multiplicity: in virtue of singular 'impulses' and hypotheses, examples that let one visualize how hypotheses are grafted to concepts, inventive forms of demonstration that let us ascertain the scaffolding's soundness. Indeed, the Peircean methodology of scientific investigation – a cycle between abduction (first), deduction (third) and induction (second) – is paradigmatically incarnated in mathematics, if we extend the planar 'cycle' to comprehend a 3-*dimensional spiral*, both recursive and amplifying.

Pursuing certain images in Grothendieck's work will help us further comprehend the amplifying spiral of mathematical creativity. From the 'cohomology deluge' **[158]** expressed in his correspondence with Serre (1956), to his musical vision of motivic cohomology **[145–146]** in *Récoltes et semailles* (1986), passing through the technical construction of the great cohomologies (1964) that would lead to the resolution of the Weil conjectures **[142, 159]**, Grothendieck sets out from a vague image (firstness, abduction: deluge), which he submits to the complex and extensive filter of definability and deduction (thirdness, deduction: étale cohomology), which he then contrasts with other invariants (secondness, induction: other cohomologies), which provokes a new vision (firstness, abduction: motifs, musicality). Note that mathematical inventiveness *is not uniquely restricted* to the realm of abduction or firstness, where the creative hypothesis obviously takes precedence, but also takes place in the realm of demonstration and the contrasting of examples. Indeed, as the Platonic *mobile base* suggests, neither invention nor discovery are absolute; they are always *correlative* to a given flow of information, be it formal, natural or cultural. It is in an antecedent transit, for instance, that Grothendieck *discovers* motifs, though a manner of representing them would later be *invented* by Voevodsky **[147, 259]**. In a similar manner, Zilber *discovers* the emergence of 'groups everywhere' **[256]**, hidden in the theory of models, though only later, together with Hrushovski, would he *invent* the Zariski geometries **[256]** by which the ubiquity of groups could be represented. There are a profusion of other examples, all of which seem to be governed by an initial, elementary typology: a *perception/vision/imagination of a generic situation*, associated with a broad spectrum of applicability (the realm of discovery), which is *interlaced* (without determinations of priority regarding the *direction* of the interlacing) with a *construction/framework/realization of many particular concretizations* in the adopted spectrum of applicability (the realm of invention). In this way, we once again find the One (discovery) interlaced with the Many (invention), through an abstract integral and differential calculus that is both mobile and complex.

The *recursive modal transit* between the possible, the actual and the necessary is one of mathematical creativity's greatest strengths. The *conjunction* of the three emphasized terms ('transitoriness', 'recursivity', 'modality') to some extent explicates the specificity of mathematical thought. On the one hand, we have seen how, throughout the twentieth century – with the works of Gödel, Grothendieck, Lawvere, Shelah, Zilber or Gromov, among many others – mathematics has opened the vital floodgates to the *relative*, while always searching behind that movement for the proper *invariants*: the transitoriness of (quasi-)objects and processes is recognized, but certain ubiquitous modes are sought in the flux (an epistemological leap from the 'what?' to the '*how?*'). On the other hand, we have also seen how the hierarchization of mathematics involves incessant processes of *self-reference*, which yield knowledge by recursion of the (quasi-)objects and processes at stake: knowledge distributes itself in layers and strata (mathematics as architectonics), and the interrelation of local information hints at a global vision of mathematical 'entities'. Finally, we have observed how *free* combinations (firsts), in the abstract and the possible, contrast with facts (seconds) of the physical world, and help flesh out a comprehension (third) of the cosmos as a whole: the weavings between mathematics and physics have been perennial, and have once again reached startling heights (Arnold, Atiyah, Lax, Witten, Connes, Kontsevich). It is between the inventive *freedom* of concepts and the inductive and deductive *restrictions* of calculation that mathematics has situated itself.

The work of Jean-Yves Girard is emblematic in this sense.[385] His monumental *Locus Solum*[386] breaks with our received ideas and completely *upends* our understanding of logic. Beyond syntax and semantics, Girard pierces through to the *geometry of logic's rules* and extensively explores logic's *locative transformations*. As a substitute for proofs and models, Girard introduces the concept of 'design' as the 'locative structure of a proof in the sequent calculus', and elaborates a subtle and complex

---

385 We are grateful to Zachary Luke Fraser for having indicated the pertinence of Girard's works to us in the context of ideas of 'flow' and 'obstruction', in both a logico-geometrical sense (*Locus Solum*) and in a methodological sense (*Du pourquoi au comment*).

386 J.-Y. Girard, '*Locus Solum*: From the rules of logic to the logic of rules', *Mathematical Structures in Computer Science* 11, 2001: 301–506.

instrumentarium for the combinatorial and geometrical transformations of designs (processes of normalization, theorems for addition, multiplication and quantification, soundness and completeness for nets). The de-ontologization of the enterprise – or, better, its new collocation in a *transitory ontology* – is remarkable. 'Designs' require no material substrates, whether syntactic or semantic, to be understood. Their pragmatic behavior, the abstract disengagement of their flows and obstructions, and their categorical manipulation are entirely sufficient.[387] *In tune* with the search for an *initial protogeometry* – having this in common, as we have seen, with the works of Zilber, Freyd and Petitot – Girard situates his methodological reflection in a leap from the 'why?' to the 'how?'[388] The French logician situates the lambda calculus, denotational semantics and his own linear logic in relation to this 'how'. The underlying substrate of the 'what' is not only unnecessary, but may even become noxious. The *liberation* of objects, the power to *look at* rules categorically and handle them in terms of a pragmatico-geometrical 'how', does indeed constitute,

---

387 Girard indicates three steps in the development of logic: 1900–1930, 'the time of illusions'; 1930–1970, 'the time of codings'; 1970–2000, 'the time of categories'. Girard's own work can be seen as attempting to cross the threshold of this third era, and bears the deep influences of category theory's mathematical development in an effort to exceed its reach. While category theory reveals the *functional*, *structural*, and *universal* dimensions of logical proofs (what Girard calls their 'spiritual' aspect), Girard's 'ludics' and 'geometry of interaction' aim to uncover their *interactive*, *dynamic* and *singular* side (what he calls their 'locative' aspect).

388 J.-Y. Girard, 'Du pourquoi au comment: la théorie de la démonstration de 1950 à nos jours,' in Pier, *Development of Mathematic*, 515–45

as we have seen throughout this book, one of the highest points of contemporary mathematical creativity.

In that environment of flows and obstructions, a general thinking of *residues* and *sedimentations* comes to be of great assistance for the better understanding of the processes of genesis at work in mathematics. Merleau-Ponty proposes that a science of 'sedimentations' be founded,[389] whereby the circle of man and nature would be closed through an *operative* body that interlaces the visible and the invisible – a body, that is to say, that conjoins the horizon of a general world (the visible) with the horizon of an underworld (what is seen by the seer), that is *inserted* in turn into the first horizon. Knowledge, rooted in a body, but mixed with a web of world horizons, refuses the Cartesian caesura of mind and world, and connects knowing and nature in a *continuous* fashion. Nothing happens 'outside the world', the senses, or vision in particular. Cultural horizons, the interpreter's contextualizations, and sedimentations are vital for knowledge, and for mathematics in particular, which turns out to be profoundly human. Phenomenology interlaces the human eye, the general horizons of the world into which vision is inserted, and the particular subhorizons where 'things' are reborn through the body of the observer. And so, on perceptual grounds, the sediments of culture, knowledge,

389 Merleau-Ponty, *Notes des cours...*, 44.

and social life continue to accumulate, and 'things' are modalized through multiple horizons, whereby we detect them in their various registers (*structure*, *fibration*, *function*, *sensation*, etc.). Figure and ground likewise outstrip their dualization, as they interlace with each other in an incisive and visible continuum, not only in the modern manifestations of art (Mallarmé, Proust, and Cézanne, as reread by Merleau-Ponty), but also, as we have seen, in the contemporary manifestations of mathematics.

In *Eye and Mind*, Merleau-Ponty describes the body operative in the domains of knowing as a 'sheaf of functions interlacing vision and movement' [151]. As we have been indicating, that sheaf serves as an *interchange* (à la Serres) between the real and the imaginary, between discovery and invention, and allows us to capture the *continuous transformation of an image into its obverse*, through the various visions of interpreters. Two of the major theses of the late Merleau-Ponty combine the necessity of both thinking the *recto/verso* dialectic and thinking in a continuous fashion:

1. What is proper to the visible is to possess a fold of invisibility, in the strict sense.
2. *To unfold the world without separating thought is, precisely, modern ontology.*[390]

390 Ibid., 85, 22.

As we have seen, many constructions specific to contemporary mathematics allow us to corroborate, with all the desired technical support, both of Merleau-Ponty's theses. The 'fold of invisibility' is particularly striking in what we have called the 'structural impurity of arithmetic' **[41]**, where the most important signposts leading to the resolution of Fermat's Theorem **[187]** are literally *invisible* from the discrete perspective of the natural numbers, until we pass through to the *obverse side* of the complex plane. Likewise, to 'unfold' mathematics, without separating its subregions, is one of contemporary mathematics' major strengths, and underlies, in particular, the exceptional richness of Grothendieck's thought, the constant transgressor of artificial barriers and explorer of *natural continuous connections* between apparently disparate images, concepts, techniques, examples, definitions and theorems.

The 'manner' **[156]**, or style, through which the great mathematicians produce their works is another problematic that analytic philosophy of mathematics intrinsically neglects. The works of Javier de Lorenzo **[76–78]** opened up an important seam at this point, several decades ago, but one which has nevertheless not been sufficiently mined. In this work's second part, we have described a few concrete registers of *local ways* of doing mathematics, ways we have not analyzed (and cannot do so: it is work for another essay) from the viewpoint of the style's *global configuration*. Nevertheless, from the phenomenological perspective on creativity adopted here, we can specify certain forms of overlapping and sedimentation, which, *counterpointing* **[206]** forms of break and rupture, help us delineate the *stylistic spectrum* of the creators of mathematics. A basic observation, in fact, shows us that the mathematical creator proceeds by way of gradual exercises of *weaving* (or 'back-and-forth') between generic (powerful and vague) images and various successive restrictions (definitions, theorems, examples) that allow her to go on sharpening the original 'impulses' or intuitions. The *counterpoint* between the deductive overlapping and the imaginal rupture is thus *necessary* in the first steps of creation, although, later, sedimentation tends to overwhelm the break.

That counterpointed textile involves peculiar modes of enlacing and correlating. Perhaps close to music in this sense, mathematics discovers *both* symmetries/harmonies *and* ruptures of symmetry/harmony, which it must then distill by way of many developments, variations, and modulations, in well-defined languages that allow it to go on to 'embody' the great harmonies or ruptures, whether seen or at first 'hidden' (remember Grothendieck, and his listening to 'the voice of things' **[151, 159]**). *Reason-ability* (= reason[a][sensi]bility, in Vaz Ferreira's sense) is vital here, and we should literally *glue together* a free, diagrammatic, imaginal sensibility (the realm of the *aesthetic*), and a normative, ordering, structuring reason (the realm of the *ethical*), whether in lengthy disquisitions on the pentagram or long series of mathematical deductions. But the very *polar coherence* of mathematical creativity (which, by contrast, does *not* occur in musical creation) obliges us to observe the situation's *underside*, *counterpointing it* with an inventive and liberating 'aesthetic reason', and with, above all, a contrastive and communal 'ethical sensibility'. Mathematics thus succeeds in transcending the imagination, unreined by the isolated individual, and becomes the greatest imaginary construction of which an entire *community* is capable.

The accelerations and decelerations in the processes of mathematically 'gluing' together conjectural images, partial hypotheses, imaginary residues, real examples and theorematic sedimentations cover the most diverse situations possible. In many eidal ascents deductive stratifications and generic visions may take precedence, over and above secondary contrastings (residues, examples): this is the case, for example, in Grothendieck-Dieudonné's EGA **[166–169]**, in Lawvere's conception of a set theory without local Von Neumann elements **[195]**, in the beginnings of Langlands's functorial program **[180]**, or in Shelah's first glimpses of the nonstructure theorems **[196]**. In turn, in what we have called quiddital descents, the richness of residues, obstructions and examples (differential equations à la Atiyah or Lax, noncommutative or quantum groups à la Connes or Kontsevich), tends to take precedence over 'primordial' images, generic instrumentaria, or deductive machinery. In the archeal findings that we have pointed out (such as the 'geometrical nuclei' in Zilber or Gromov, or the 'logical nuclei' in Freyd or Simpson), we do *not* observe a particular directionality or predominance in the combination of images, hypotheses, residues, examples, definitions and theorems that develop the mathematical archetypes which correlatively 'dominate' certain classes of structures. In the latter, in fact, particularly in the works of Zilber and Gromov, the *incessant weaving* between the most concrete and most abstract is not only *noncircumventable*, but constitutes something like a genuinely original, systematically oscillatory, *manner*, one we could perhaps consider proper to the Russian 'school'.

Going back to the work of Nicholas of Cusa, Hans Blumenberg recalls that the world divides into '*visibilia, invisibilium imagines* [visible "things", images of the invisible]', and how the universe of '*mathematicalia* [mathematical "things"]' allows us to refine our vision from a clear 'methodological vantage point: the availability to freely perform variations, the possibility to experiment while freely establishing conditions'.[391] We have already

---

391 H. Blumenberg, *Paradigmas para una metaforología* (1960) (Madrid: Trotta, 2003), 242–3. (Originally published in German as *Paradigmen zu einer Metaphorologie*, in *Archiv für Begriffsgeschichte* [Bonn: 1960]. [Tr. R. Savage as *Paradigms for a Metaphorology* (Ithaca: Cornell University Press, 2010)].

emphasized the fundamental *freedom* of mathematics, its variational richness in the realm of *possibilia* and its capacity to enter the dialectic of *visibilia* and *invisibilium*. To some extent, that plasticity is due to another profound counterpoint in mathematical creativity: the pendular weaving between the *metaphorical* and the technical. Mathematics has not only been a producer of *external* metaphors for the unfolding of thought, but often proceeds *internally*, in its processes of creation, by way of vague and potent metaphors, still remote from precise, technical delimitations. On this note, it is curious that analytic philosophy of mathematics should have proscribed the study of *mathematical metaphorics* as something unbefitting of exact knowledge, when its entire program emerged from explicit metaphors – metaphors later enshrined as *myths*, has Rota has pointed out **[86]**: atomism, absolute, dualism, foundation, truth, etc.

A phenomenology of mathematical metaphorics should draw on Blumenberg's extensive work[392] in the historical tracking of metaphors in philosophical language. Starting from the fundamental (Husserlian) antinomy

---

392 Major monographs: *Paradigmen zu einer Metaphorologie* (1960) (*Paradigms for a Metaphorology* [2010]), *Die Legitimität der Neuzeit* (1966) (*The Legitimacy of the Modern Age* [1985]), *Die Genesis der kopernikanischen Welt* (1981) (*The Genesis of the Copernican World* [1987]), *Die Lesbarkeit der Welt* (1979), *Lebenszeit und Weltzeit* (1986), *Höhlenausgänge* (1989). Minor essays (and gems): *Schiffbruch mit Zuschauer. Paradigma einer Daseinsmetapher* (1979) (*Shipwreck with Spectator: Paradigm of a Metaphor for Existence* [1996]), *Das Lachen der Thrakerin. Eine Urgeschichte der Theorie* (1987), *Die Sorge geht über den Fluß* (1987) (*Care Crosses the River* [2010]), *Matthäuspassion* (1988). A good presentation of Blumenberg's work can be found in F. J. Wetz, *Hans Blumenberg. La modernidad y sus metáforas* (Valencia: Edicions Alfons el Magnànim, 1996).

between the *infinitude* of the philosophico-scientific task and the *finitude* of human individualities, Blumenberg studies the complex lattice of graftings and ruptures between the 'time of the world' and the 'time of life', where mathematics plays the role of an exception. Reviewing Husserl's *Logic* of 1929, Blumenberg emphasizes how the master introduces the metaphor of a '*historical sedimentation as the product of a static ideality of dynamic origin*'.[393] The metaphor's *apparent* contradictoriness stems from disparate appropriations of Plato (a static reading *versus* Natorp's neo-Kantian, dynamic reading, taken up again by Husserl), but superbly recollects several of the primordial modes of mathematical creativity that we have been emphasizing. Creativity proceeds within the 'time of life', within what Merleau-Ponty called the operative body, but is always extending itself to the 'time of the world' that envelops it; sedimentation is therefore historical, and takes place through a web of ideal positions that appear static, but which emerge on a ground of dynamic, polar mediations. Indeed, mathematics – understood as the study of the exact transits of knowledge ('dynamics') – constructs partial invariants ('static ideality') in order to gauge the obstructions and osmoses in transit, and then gathers the various registers obtained ('historical sedimentation').

---

393 H. Blumenberg, *Tempo della vita e tempo del mondo* (Bologna: il Mulino, 1996), 391.

In the incessant mediation between dynamics and stat-
ics, reality and ideality, world and life, the *Fundierung*
according to Rota (who also, in his own way, takes up
and reinterprets Husserl's ideas) combines two central
processes in the construction of mathematical webs:
facticity and *functionality*.[394] For Rota, a mathematical
(quasi-)object – through its many forms, from the vague
metaphorical image to the carefully defined and delimited
technical (sub-)object, passing through intermediate
modes, examples and lemmata – should be, on the one
hand, pragmatically inserted into a context (facticity),
and, on the other, correlatively contrasted in that context
(functionality). Mathematics, which lives *synthetically*
on both the factical level (contextualization) and the
functional level (correlation), appears irreducible to any
supposed 'objectuality'. The *Fundierung* studies how mathematical *processes* lace into one another, *independently* of
their analytical classification (via $\in$ or $\subseteq$). The processes
(or quasi-objects) are understood as *poles* of a relation
of *stratification*, with complex gradations between them.
Mathematics studies the transformation and splicing of
those various gradations, *independently* of any illusory
ultimate 'ground' on which they would rest. Beyond
certain supposedly stable and well-founded mathematical 'objects', the *factical and functional webs of coupling*

---

394 Palombi, *La stella e l'intero…*, 62.

between mathematical quasi-objects, in the broadest possible sense (metaphors, ideas, processes, conjectures, examples, definitions, and theorems), thus constitute the true spectrum of mathematical phenomena.

A phenomenology that seeks to capture the transits of mathematics in an adequately faithful and nonreductionist manner must take many polarities into account: analytic decompositions and synthetic recompositions, modes of differentiation and integration, processes of localization and globalization, particularities and universalities, and forms of creativity and discovery, among others. As we have seen in the preceding chapters, the differential-integral back-and-forth is not only situated on the epistemological level ('how'), but *continuously extended* into the 'what' and the 'where' of the quasi-objects at stake. The many strata/environments/contexts of mathematics, according to Rota's *Fundierung*, respond to a complex regimentation of the prefix *trans-*, on ontic and epistemic levels, breaking the customary barriers of philosophical reflection.

For a *minimal conceptualization* of the *trans-*, a consideration of the following operations is required:

(A) Polarities (or 'adjunctions'):
- decomposition/composition
- differentiation/integration
- deiteration/iteration
- particularization/universalization
- localization/globalization
- residuation/potentiation

(B) Mediations
- oscillation
- mixing
- triadization
- modalization
- sheafification

Figure 15. *Polarities and mediations in a general operativity of the* trans-.

Analytic philosophy of mathematics excludes these mediations from its realms of investigation, which perhaps explains its incapacity to capture the universe of advanced mathematics. The first dual pair ('decomposition/composition') recalls the *necessary and irreducible dialectic* between analysis and synthesis, a dialectic that runs through the entire history of philosophy and mathematics. In turn, the first mediation ('oscillation') recalls a *necessary and irreducible* pendular variation of thought, forever stretched between opposing polarities. The second mediation ('mixing' à la Lautman) accompanies that inevitable pendular oscillation with the awareness that we must construct mixtures to serve as support structures for an extended reason ('reasonability'), mixture being understood here in the sense of *sunthesis* (as composition, which is reversible), as opposed to *sunchusis* (as fusion, which is usually irreversible). The second dual pair ('differentiation/integration') recalls one of the major, originary problematics of philosophical and mathematical thought: the dialectic of the Many and the One. The third dual pair ('deiteration/ iteration'), together with the third and fourth mediations ('triadization' and 'modalization'), constitutes one of the great richnesses of advanced mathematics, something that is usually *invisible* in elementary strata or from analytic perspectives, but which, by contrast, is reflected with great acumen in the original operative nucleus of the Peircean architectonic [111–121]. Indeed, the Peircean emphasis on the rules of deiteration/iteration represents one of Peirce's most profound contributions, whether from a logical point of view (the rules codify the definitions of the connectives), or from a cognitive one (the rules codify *creative* transfers of information). Similarly, the systematic Peircean triadization and its modal filtration secures the plural richness of the pragmaticist architectonic.

The fourth and fifth dual pairs ('particularization/universalization' and 'localization/globalization'), together with the fifth mediation (sheafification), answer more specifically to contemporary mathematical forms of thought. In certain well-delimited cases, sheafification allows local information to be glued together in a coherent fashion, and takes us to global quasi-objects that capture the transit of the information in the sheaf's fibers. Mathematics, then, is to a large extent concerned with calibrating the calculable osmoses and obstructions in those back-and-forths between local and global properties, in the realms of space, number, structure and form. The sixth dual pair ('residuation/potentiation'), for its part, captures the richness of certain strata of mathematics with strong coherence properties (like the calculus of the complex variable, or topos theory), where the residues become (quasi-)objects that are completely *reflective* of their environment, and exponents of mathematics' visionary possibilities.

In all of these processes there is an enrichment of the Peircean *summum bonum* – understood as the 'continuous increase of potentiality'[395] – and an explosion of mathematical creativity into the most diverse forms. The *metaphorical and analogical webs* studied by Châtelet [93–95] combine with precise modes of sedimentation – examples, definitions, theorems – giving rise to a sophisticated 'integral and differential calculus' of mediations and gradations (a *Fundierung*) that orients the evolution of mathematical thought. Demonstrative

---

395 For an excellent presentation of the place that a broad 'reasonability' occupies in Peirce's system, oriented by the *summum bonum*, and for a study of its correlations with sensibility, creativity, and action, see S. Barrena, *La creatividad en Charles S. Peirce: abducción y razonabilidad*, doctoral thesis, Department of Philosophy, University of Navarra, Pamplona, 2003 (part of which has been published as S. Barrena, *La razón creativa. Crecimiento y finalidad del ser humano según C.S. Peirce* [Madrid: Rialp, 2007]). Beyond Peirce himself, Barrena's works represent a remarkable contribution to the understanding of creativity in general.

'rationality' is wedded to imaginary 'reasonability': in the processes of articulating/mediating/gluing the various 'dual pairs' we have just mentioned, lies the discipline's inventive malleability. What we are dealing with here is a peculiar plasticity, which combines a joyous capacity for movement (a facility of transit through the world of the possible) and a careful handling of dynamic differentials (a control of change in the realm of exactitude). It is perhaps in this *connection between plasticity and exactitude* that the true strength of mathematics is found, a connection that is crucial for the processes of mathematical *creation*. Any perspective that neglects that plastic '*art of doing*' **[327]** neglects the *living nucleus* of the discipline itself.

Serre, who should no doubt be considered as one of the major *stylists* of contemporary mathematical literature, used to point out the importance of mixtures in mathematical creativity, and of the presence of a vital inventive penumbra – 'I work at night (in half-sleep), [which] makes changing topics easier' **[177]** – behind the supposed luminosity of proof. It could be said, observing the almost *crystallographic* luster of Serre's own work, that the great mathematical creator laces his perpetual decantation of the penumbra (the realm of discovery) with a rare capacity to reveal/construct luminous crystals (the realm of invention), on his zigzagging path. Indeed, it is remarkable that Serre's limpid style, astonishingly smooth and 'minimal', should, in the author's own words, reveal itself as a 'wonderful mélange' situated on penumbral ground. In the same sense, many of contemporary mathematics' *crystallographic gems* emerge from the *obscure grounds* where they are born: Grothendieck's motifs with their musicality in major and minor keys [145–146], Langlands's letter to Weil with its putative unseriousness and casual tone [180–182], Atiyah's Index Theorem with its excavation of murky depths [208], Cartier's dream with its shifting terrains in mathematical physics [226], Zilber's extended alternative with its obscure intuition of the complex variable's logical behavior [257], and Gromov's *h*-principle with its ground of discordant situations in the penumbra of differential equations [263], amid many other examples that we took up in part 2 of this book.

As we have seen throughout this chapter, advanced mathematical creativity can be understood only through perspectives that *reflect* the phenomenology of nontrivialized mathematical transits: intertwinings in webs, nondualist gradations, contaminations on a continuum, recursive modal interlacings, dialectics of sedimentations and residues, partial osmoses between metaphorical images and technical objects, global gluings over coherent local adumbrations, factical and functional processes of intermingling (*Fundierung*), and systematic mediations between polarities. In the realm of elementary mathematics, these phenomena tend to vanish, owing to the reduced complexity of the entities in question. On the other hand, from the perspectives of analytic philosophy (and regardless of the realm observed, whether elementary or advanced), these phenomena are also neglected, since they are usually considered 'ill-defined', or impossible to define at all (we hope to have shown in chapters 8 and 9 that this is *not* the case). Perhaps these two tendencies, together with the tendency (predominant in the philosophy of mathematics) to study the elementary from analytic perspectives may explain the scant concern that has, until now, been given to the problematic of mathematical creativity.

CHAPTER 11

MATHEMATICS AND CULTURAL CIRCULATION

In this final chapter we will take up two remaining considerations, with which we will try to round off our work. If, in chapters 8–10, we have emphasized a turn to questions connected with the '*how?*' in contemporary mathematics, and if, in chapters 4–7, we have reviewed a few emergences ('where?') of precise and delimited problematics ('why?') in detail, in this chapter we will study, on the one hand, the *general* interlacing by which contemporary mathematics (1950 to today) can be distinguished from prior mathematics (a *conceptual* 'what/how/why' *synchrony*, beyond a merely diachronic cut happening around 1950). And, on the other hand, we will also examine the *general* positioning of mathematical thought within culture, and the ways in which it naturally shares a frontier with aesthetics (a *conceptual geography* of the 'where?').

In chapter 1 we distinguished from a 'bird's eye' view (a perspective both distant and evanescent) some features that allow us to provisionally separate modern mathematics (from Galois to about 1950) from contemporary mathematics (from around 1950 to today). We will summarize these characteristics in the following table (with 1–5 being implicit in Lautman's work **[30]**, and 6–10 being implicit in the developments of contemporary mathematics [41]), and then explicate the *synchronic conceptual grounds* that may help to explain characteristics 1–5 and 6–10, as well as the diachronic cuts situated around 1830 (the beginnings of modern mathematics) and 1950 (the beginnings of contemporary mathematics).

In what follows, in the first part of this chapter, we will study some forms of *internal circulation* within the conceptual realm of mathematics, which will help us to intrinsically distinguish certain periods of mathematical production; in the second part, we will study some forms of *external circulation* in the general realm of culture, which will help us better explicate certain modes of mathematical creativity through suitable correlations and contrasts. In an exercise of *counterpointing*, we will go on to pinpoint the 'why' of modern mathematics' emergence and the 'why' of its later evolution into the questions, methods and ideas of contemporary mathematics. We will discuss the characteristics observed by Lautman (1–5), with regard to both their positive side and their *obverse*. This will lead in a *natural* way to characteristics (6–10), which reflect certain crucial features of the new conceptual grounds at stake in contemporary mathematics.

	MATHEMATICS	
	MODERN (1830–1950)	CONTEMPORARY (1950–present)
1. *complex hierarchization* systems of mediations	✓	✓
2. *semantic richness* irreducibility to grammars	✓	✓
3. *structural unity* multiple polarities	✓	✓
4. *dynamics* movements of liberation/saturation	✓	✓
5. *theorematic mixing* ascents and descents	✓	✓
	transits/obstruction hierarchies/structures modeling/mixing	
6. *structural impurity* the arithmetical via the continuous		✓
7. *ubiquitous geometrization* archeal geometric nuclei		✓
8. *schematization* categorical characterizations		✓
9. *fluxion and deformation* obverse of usual properties		✓
10. *reflexivity* complex forms of self-reference		✓
		fluxions/alternations schemes/nuclei reflection/sheafification

Figure 16. *Some conceptual features that help demarcate modern and contemporary mathematics.*

Modern mathematics more or less surfaced through the work of Galois and Riemann, with the introduction of *qualitative* instruments for the control of quantitative problematics, from a point of view that is both positive (transits) and *negative* (obstructions). On the one hand, in fact, the structural interlacing of the hierarchy of Galois subgroups and field extensions allow us to control the behavior of roots of equations (and allow us to ascertain, among other things, the general *unsolvability* of quintic equations); the topological properties of Riemann surfaces, on the other hand, allow us to control the ramification of multivalent functions of complex variables (and allow us to ascertain the *inequivalence* of surfaces like the sphere and the torus, for example). From its very beginnings, modern mathematics has faced a clearly definable, generic problematic: (*A*) *the study of the transits and obstructions of mathematical objects, with qualitative instruments, associated with structural mediations and hierarchies.* The conceptual ground at stake is partially reflected in characteristics 1–5 mentioned above, and corresponds to a genuinely fresh and novel focus of mathematical perception, one which has since *systematically* opened itself to a qualitative comprehension of phenomena and a reflexive understanding of the very *limitations* (negation, obverse, obstruction) of that partial comprehension.

The *complex hierarchization* of mathematics (1) due to Galois and Riemann yields many of the richest branches of modern mathematics (abstract algebra, functional analysis, general topology, etc.), but it is a process (or, better, a collection of processes) that naturally tends toward saturation on each level of the structural hierarchies in question. For example, behind the profusion of semigroups and groups, contemporary mathematics would perform a *schematization* (8) that would open the gates to groupoids in general topoi, or, even more schematically, to *operads* [235]. On the one hand, the notorious *structural unity* (3) of modern mathematics (a unitary solidity that is perhaps its most distinctive feature) is, in contemporary mathematics, counterpointed by a sophisticated *extension* of the unitary toward that dialectical unity's polar frontiers, with an altogether original capacity [201, 231, 264, etc.] for tackling *fluxions and deformations* (11) of structures which, it once seemed, could be understood only rigidly. On the other hand, the remarkable *semantic richness* (2) of modern mathematics, with the enormous multiplicity of models arising in the period between 1870 and 1930, in all the realms of mathematical action (geometries, sets, algebras, functional spaces, topologies, etc.), would later yield a *reflexive* vision (10) of that diversity, in tandem with the construction of the instruments needed for a reintegration of the local/differential into the global/integral.

Sheaf theory, whose construction can be traced to the period between 1943 and 1951 (a conceptual synthesis represented in figure 13 **[163]**, and whose diachronic emergence we have already observed **[163, 285]**), constitutes, for us, the *decisive index* that allows us to capture the changes/delimitations between modern and contemporary mathematics. Indeed, *in both their conceptual ground and their technique*, sheaves effectively symbolize one of the great general problematics of mathematics in recent decades: (*B*) *the study of fluxions and deformations (arithmetico-continuous, nonclassical) of mathematical quasi-objects, with instruments of transference/blockage between the local and the global, associated with processes of schematization and self-reference.* Of course, this problematic, which we have surveyed from above, neglects other important applied, calculative and computational aspects of contemporary mathematics. It is nevertheless clear that, to a large extent, it includes aspects of the major mathematical works of the second half of the twentieth century, and central aspects of *all* of the works that we have carefully reviewed in chapters 4–7, in particular.

Contemporary mathematics thus inscribes itself in a fully *bimodal* spectrum, in Petitot's sense: *simultaneously* physical and morphologico-structural. Indeed, as we saw with Rota, the incarnation of mathematics' quasi-objects is *at once* factical and functional **[346]**. Lived experience and knowledge occur in relative environments of information transformation, and not on absolute foundations. Mathematical *intelligence* thus consists in modes of knowledge-processing that *lead from in/formation to trans/formation*, modes including both the *analytic* dismemberment of information and the *synthetic* recomposition of the representations obtained in correlative horizons. The bimodal and the bipolar, which yield progressive gradations and precise *frontier conditions* in transit, lead to the natural mediations/mixtures characteristic of mathematical knowledge. In the incessant search to precisely and correctly determine the multiple boundaries of mathematical quasi-objects, the great problematics of modern and contemporary mathematics, (A) and (B), answer to conceptual grounds that, with respect to certain frontier conditions, are welldetermined: (A) takes up the *de/limitation* of classes of classical structures, in a first approximation that partially *fixes* certain coordinates in modern knowledge, while (B) takes up the *extra/limitation* of those classes, deforming and differentiating them (locality) in order to then reintegrate them (globality), in a second approximation that *frees* certain variations in contemporary knowledge.

The net result of the conjunction of problematics (A) and (B), the situation in which we find ourselves today, is a thorough, mathematical understanding of certain *relative universals*, that allow us to combine modern mathematics' fundamental quest for 'universality' with contemporary mathematics' modulation toward the 'relative' (and its search for the invariants behind the transit). Indeed, many of the prominent works that we have been reviewing respond in a precise and delimited manner to the constitution of *webs of relative universals*: the Langlands correspondence [181], Lawvere's 'unity-and-identity of opposites' [191], Shelah's general theory of dimension [197], Connes's noncommutative geometry [222], Kontsevich's quantum cohomology [235], Freyd's allegories and intermediate categories [243], Simpson's reverse mathematics [248], Zilber's extended trichotomy [256], and Gromov's *h*-principle [263], among many other accomplishments. In all of these cases, which should be seen as typical expressions of contemporary mathematics, we see, to begin with, a full assimilation of mathematical *dynamics*; secondly, a search for ways of controlling that movement (that is to say, the ways in which adequate *frontiers* can be defined), and thirdly, a construction of delimited and technically well-adjusted quasi-objects that, with respect to the dynamics and frontiers just mentioned, act as relative universals: the Langlands group [184], adjoint functors [191], dividing lines of the Main Gap [199], the Lax-Phillips semigroup [219], the Grothendieck-Teichmüller group [225], *Cor–Split–Map* functors [245], second order canonical subsystems [249], Zariski geometries [256], Gromov's smooth inequalities and invariants [260], etc.

From a *metaphorical* perspective, the transition between modern and contemporary mathematics corresponds to a process of *liberation* and *variational amplitude*, reflected in the internal circulation of concepts and techniques that we have been describing. A profound 'shifting of the soil' **[150, 278]** has *freed* mathematics. Through a continuous process, we have travelled a path *of reason/ imagination's progressive amplification*: the working of mathematics' 'soil' (the analytic/set-theoretic reconstruction of mathematics), the experience of the soil's 'shifting' (relative consistency proofs à la Gödel, relative mathematics à la Grothendieck), the 'bimodal' understanding of mathematical transit (the emergence of category theory, the return of close ties with physics), and the synthetico-mathematical construction of 'relative universals'. The *height of reason* allows us to contemplate the shiftings and sedimentations that together compose the terrains of contemporary mathematics. What we are dealing with here, of course, is the gestation of a new topography, whose looming presence the philosophy of mathematics must recognize without delay, and which will surely shatter its rigid academic matrices.

Although the aforementioned modulations must be inscribed on a continuum, we may note that this progressive amplification is stretched across discrete counterpointed looms. On the one hand, *theorematic mixing* (5) has led to the singular discovery of archeal geometric nuclei, in the currents of a *ubiquitous geometrization* (7) by which the modern mixings can, to a large extent, be governed from new perspectives. There is an evident tension here between continuous mediations and discrete nuclei (motifs, allegories, combinatorial groups and semi-groups, Zariski geometries, etc.), by which it seems such mediation can be controlled. On the other hand, modern mathematics' *structural unity* (3) has widened its *margins* to a possible union of *fluxion and deformation* (9) where the structural seems to be no more than (an admittedly central) *part* of a total, complex dynamic landscape; in the extension of mathematical structures' *generic regimentation*, discrete leaps (quantizations) counterpoint continuous deformations, giving a new form to Thom's aporia in the current landscape of mathematics.

Another profound transition helps to explicate the 'why' of the demarcation between modern and contemporary mathematics. The rising power of the *asymptotic* – counterpointing the fixed or the determinate in modern mathematical *logoi* – permeates many of contemporary mathematics' major forms. From the emergence of many classifier topoi and inverse limits with which we may 'glue' the classifiers together – which yields an asymptotic understanding of logic, elsewhere corroborated by Caicedo's results in sheaf logic **[283]** – to Gromov's great asymptotic sweeps through Hilbert's tree **[262]**, passing through the 'covering approximations' of what we have called the *Grothendieck transform* [320–322] or through the richness of partial coverings of the real in the quiddital approach of Atiyah, Lax, Connes or Kontsevich, contemporary mathematics has been able to express and control, with the utmost conceptual and technical power, the crucial notion of a *hiatus* between fragments of knowledge. The hiatus, understood as a cleft or a fissure – that is to say, as a '*between*' *on the obverse* of our conceptions – in turn becomes an aperture to the unexplored, just as it appears in Merleau-Ponty's *Eye and Mind* [338]. The *to ti einai* (the essential of essence) [124], cannot be described as a universal concept or 'entity', but precisely as a generic form of hiatus, inevitably present both in the world ('transitory ontology' – chapter 8) and in our approach to it ('comparative epistemology and sheafification' – chapter 9). The *partial*, *relative*, *asymptotic covering* of that hiatus is one of philosophy's primordial tasks, and one that can now be embarked upon with an entire series of concepts, instruments, methods and examples belonging to *contemporary mathematics*.

Flux, shifting, and hiatus have always and everywhere engulfed us. Novalis's vivid resurgence in contemporary culture is not a matter of chance, nor is the recognition of the contemporary relevance of Peirce's asymptotic architectonic. Nor, moreover, does it seem to be mere chance that the beginnings of contemporary mathematics can be situated around the emergence of sheaf theory, a theory as sensitive as any to the shifting covering of local obstructions. The enormous philosophical importance of *this mathematics in which we find ourselves immersed* is to a large extent rooted in its extraordinarily rich conceptual and technical arsenal, an arsenal built for an increasingly careful study of flux, shifting, and hiatus. As Corfield has stressed, we must 'not waste it' **[100]**.

* * *

The changes and advances in mathematics during the second half of the twentieth century have been remarkable. We have seen that these transformations correspond to a gradual amplification of problematics (the modulation from (*A*) to (*B*) **[356, 358]**) and an extended capacity for treating, with new technical instruments, deformations of mathematical quasi-objects, the gluing together of the local and the global, geometrical nuclei of representations, processes of self-reference and schematization, relative and asymptotic channels, and nonclassical structural graftings with physics. Behind all these accomplishments, and others still unmentioned, we may glimpse the permanent presence of a certain 'general operativity of the *trans-*' (figure 15 **[347]**), in the background of contemporary mathematics. It is interesting to observe that, as we move away from the hackneyed rupture associated with 'postmodernism' in the years 1960-1970, and approach, instead, a *web of entrances into and exits from modernity*, a *transversal traversal of the modern* – what we could call a sort of 'transmodernity' – the *internal* circulations achieved in mathematics seem to anticipate a natural *external* reflection in culture.

An apparent rift between modernity and what is unfortunately called 'postmodernity' is marked by the brilliant work of Deleuze.[396] Nevertheless, aside from the richness of 'postmodernism's' founding documents, many of the movement's subsequent, 'degenerate'[397] theses – like 'anything goes', absolute relativism, the impossibility of truth, the conjunction of the arbitrary, the dissolution of hierarchies, or the celebrated death of knowledge, among other extreme propositions – impede the *critical and comparative* exercise of reason/reasonability/imagination.

---

396 Among the immense primary and secondary literature surrounding Deleuze, let us pick out a few texts that may be useful for the philosophy of mathematics: P. Mengue, *Gilles Deleuze ou le système du multiple* (Paris: Kimé, 1994) emphasizes Deleuze's *systematic* thinking of mediations, imbrications, and fluxes, the occurrence of which *within* mathematics we have repeatedly emphasized here. L. Bouquiaux et al, *Perspective. Leibniz, Whitehead, Deleuze* (Paris: Vrin, 2006), studies the problematic of the multiplicity of points of view ('perspective'), of how they can be partially reintegrated, and how they can be used to *act* on the world. Without having drawn upon Leibniz, Whitehead, or Deleuze in our essay, we have repeatedly taken up this problematic through the Peircean architectonic, the theory of categories, and the processes of sheafification. Duffy, *Virtual Mathematics*, is a collection of articles on Deleuze's philosophy and its potential effects on the philosophy of mathematics. *Collapse: Philosophical Research and Development*, ed. R. Mackay, vol. 3, 2007 includes an important series of 'nonstandard' articles on Deleuze, which, among other things, tackle a potential 'integration' of differential Deleuzian constellations.

397 'Degenerate' should be understood in Peirce's sense of the term: being of diminished *relational complexity*. This is the case with the theses mentioned above, which *flatten* the landscape of thought.

Leaving aside the traumatic, self-proclaimed POST of their alumni, what Deleuze and Foucault teach us is how to *transit*, how to enter and exit modernity in many ways and from many perspectives. What was awoken by the second half of the twentieth century – an age of the *trans-*, if ever there was one – is better called *transmodernity*, as Rodriguez Magda had already proposed in 1987.[398]

Contemporary mathematics, which is impossible to associate with 'post'-modernity, finds a natural habitat in the *transmodernity* in which we are situated. On the one hand, contemporary mathematics is able to distinguish valences, relativize discriminately, constitute asymptotic truths, conjoin the coherent, and hierarchize knowledge, against the 'degenerate' 'post'-modern theses. On the other hand, contemporary mathematics is traversed by incessant osmoses, contaminations, syncretisms, multichronies, interlacings, pendular oscillations, coherent gluings, and emergences of relative universals, bringing it together with *transmodern* processes.[399] The richness of

---

398 R. M. Rodriguez Magda, *Transmodernidad* (Barcelona: Anthropos, 2004). Without using the term 'transmodern', the works of García Canclini and Martín-Barbero are consonant with the constant exits from and entrances into modernity that Rodríguez Magda detects. See N. García Canclini, *Culturas híbridas. Estrategias para entrar y salir de la modernidad* (1989) (México: Grijalbo, 2005), and J. Martín-Barbero, *De los medios a las mediaciones. Comunicación, cultura y hegemonía* (1987) (Bogotá: Convenio Andrés Bello, 1998). The plural richness of Latin America (and, more generally, of the Hispanic world, if we include Rodríguez Magda) has served as a natural brake on the 'postmodern' currents that, curiously, tend to *uniform* everything in difference.

399 Compare this situation with the following quotation from Rodríguez Magda, for example: 'Transmodernity extends, continues and transcends modernity; in it, certain ideas of modernity return. Some of these are among the most naïve, yet the most universal, of modern ideas – Hegelianism, utopian socialism, marxism, philosophies

contemporary mathematics, *in its overwhelming technical imagination*, disposes of a multitude of signs/operators/mediators for the subtle observation of transit. Indeed, reversing our approach, if mathematics, as has often been the case throughout history, can serve as an index for *forecasting* the tendencies of an age, then contemporary mathematics may serve as an *introduction* to the transmodernity that is now enveloping it. Just as the Renaissance may have been encrypted in Leonardo's perspectival and geometrical machines, as the baroque corresponds to Leibniz's differential and integral calculus, as classicism was forecast in Euler's serial manipulations, or as Modernity is just a temple for the visions of Galois and Riemann, Transmodernity may be encrypted in the technical plasticity of contemporary mathematics, symbolized, in turn, in the figure of Grothendieck.

If these associations or 'predictions' turn out to be more or less correct, we should nevertheless observe that, unlike the periodizations of the history of art that have been used in recent centuries – Renaissance, Baroque,

of suspicion, the critical schools [...] all exhibit this naivety. Through the crises of those tendencies, we look back on the enlightenment project as the general and most commodious frame through which to choose our present. But it is a distanced and ironic return, one that accepts that it is a useful fiction. [...] The contemporary zone is transited by every tendency, every memory, every possibility; transcendent and apparent at once, willingly syncretic in its 'multichrony'. [...] Transmodernity is the postmodern without its innocent rupturism (...). Transmodernity is an image, a series, a baroque of fugue and self-reference, a catastrophe, a twist, a fractal and inane reiteration, an entropy of the obese, a livid inflation of information, an aesthetics of the replete and its fatal entropic disappearance. The key to it is not the 'post-', or rupture, but the glassy transubstantiation of paradigms. These are worlds that penetrate one another and turn into soap bubbles or images on a screen'. Rodríguez Magda, *Transmodernidad*, 8–9.

Classicism, Modernism, and Contemporaneity, epochs in which the forms of artistic expression are *all* equally complex – the 'epochs' of mathematics are tied, by contrast, to a very palpable and *continuous augmentation of complexity* throughout their historical evolution. In fact, 'advanced' mathematics, which we have defined so as to include mathematical technique from Classicism onward **[26–28]**, clearly increases in complexity over the centuries. No doubt, the *steadily increasing difficulty of accessing* those forms of mathematical expression, in the degree it has reached today, blocks a thorough comprehension of mathematics *in its entirety*, from the most elementary to the most advanced. The task of a serious philosophy of mathematics should nevertheless be to break down these barriers, and develop a conscientious reflection on modern and contemporary mathematics. The work of an art critic who is unaware of everything that has happened in art over the last hundred and fifty years would be considered quite poor. The situation should not be any different in the philosophy of *mathematics*, and the easy complacency of reflecting on *logic* alone (in the style of the *Oxford Handbook of Philosophy of Mathematics and Logic* **[101–107]**) should be shaken off at once.

Aside from the coincidences that may be observed regarding epochal *delimitations* in the arts and mathematics, the proximity of the *creative grounds* of the artistic and the mathematical has been emphasized throughout the history of culture. Pierre Francastel, the great critic and historian of art, has forcefully pointed out how mathematics and art should be understood as the *major poles of human thought*.[400] Behind those modes of knowledge, Francastel, for his part, has observed the emergence of systems and *creative webs* wherein the real and the ideal, the concrete and the abstract, and the rational and the sensible are combined.[401]

In Peirce's triadic classification of the sciences, mathematics are situated on the first branch (*1*), in the realm of constructions of possibility. Aesthetics appears in philosophy (*2*), where, in the normative sciences (*2*), it occupies a primary place (that is to say, in branch *2.2.1*). 'Art' as such does not enter the realm of the sciences and is not to be found in the Peircean classification, but it may be

---

400 'Art and mathematics are the two poles of all logical thinking, humanity's major modes of thought', P. Francastel, *La realidad figurativa* (1965) (Barcelona: Paídos, 1988), vol. 1, 24.

401 'From the moment we accept the idea that mathematical and artistic signs respond to intellectualized knowledge and not only to simple sense data immersed in matter, we also admit the intervention of a logic, of a system, and notions of order and combination, equivalence, relation, operation and transposition appear before us. [...] Just as mathematics combines schemes of representation and prediction, in which the real is associated with the imaginary, so the artist brings elements of representation into confrontation with others that proceed from a problematic of the imagination. In both cases, the dynamism of a thought that becomes conscious of itself by expressing and materializing itself in signs interlaces, overtakes, and envelops the elements of experience and those of the logic of the mind. [...] Just like art, mathematics possesses the dualistic character by which they both reach the heights of abstraction, while remaining anchored in the real. It is in virtue of this that both mathematical and plastic symbolism preserve their operative character', ibid., 125-6. The 'dualistic' character remarked upon by Francastel should be understood as the process of a dual *intermingling* of the real and the imaginary, over a *relay* **[60]** of interlacing signs. The mediation of the relay imposes itself on the dualism of the positive and the negative, of the greater and the lesser.

seen as being very close to a *creative materiality* of type
3.2.2 or *3.2.3* – material mediations for engendering sense
(classical art) or action (contemporary art). In the Peircean
classification of the forms of knowing (and understanding
art here as a *vital* part of knowing, something we do not
find in Peirce), mathematics and art likewise emerge as
clear *polarities* (*1* versus *3.2.3*). A vision of the tree *from
below* thus provides us with a possible transit between
mathematics and art. If, metaphorically speaking, we
situate the tree on the page of assertion of the Peircean
existential graphs and view it from its *recto* (alpha graphs)
or from its *verso* (gamma graphs), we may transit between
the various realms of creation, with *interstitial* passages
between mathematics and art (dotted gamma cuts, singular points of ramification), but with blockages between
them as well (strict alpha cuts, restrictive delimitations).

The metaphorics of the tree and the graphs has a much
greater depth than may appear at first sight. The tree,
on the one hand, as a triadic *tissue*, refers us to processes
of *iterative construction* in culture, which *unfold* in time
and space. On the other hand, the graphs, as specular
images, refer us to processes of *singular vision*, *folding back
on themselves* and encoding information. In the weaving
between iteration and deiteration (which are the *main*
logical rules of the Peircean existential graphs), culture is
unfolded and refolded, in a permanent dialogue with its
main modes of creation (arts and mathematics, according

to Francastel). The creative proximities between art and mathematics, evident in the *emergence* of inventiveness, reinforce one another in the general modes of knowing, from a point of view that is formal, dual and latticial.

A proximity is nevertheless far from an identification. The creative forms of mathematics and art retain their differential specificities, and, though the polarities form a remarkable space of mediations (like the two poles of an electromagnetic field – as Châtelet reminds us **[308]**), they must begin, first and foremost, by folding back upon one another. The demonstrative realm of mathematics, cumulative and architectonic (*3*), naturally repels the intuitive, destructive and visionary (*1.2*) realm of art. In this manner, though the *modes* of creation in both realms are in close proximity, the quasi-objects at stake are extremely distinct. We are therefore faced with a very interesting *asymptotic geometry* between mathematics and art: *orthogonal* 'what?', *dual* 'how?', and *inverse* 'why?'.

The breath of aesthetics permeates mathematical creativity on at least two levels, as *detonator* and as *regulator*. Referring to the artistic imagination, Valéry writes in his *Cahiers*: 'Imagination (arbitrary construction) is possible only if it's not forced. Its true name is *deformation of the memory of sensations*',[402] and referring to the imagination in general, he speaks of the 'imaginary magnitudes – *SPECIAL*

---

402 P. Valéry, *Cahiers 1894–1914* (Paris: Gallimard, 1990), vol. 3, 219 (Valéry's emphasis).

efforts when there are displacements or tensions'.[403] We
have seen how contemporary mathematics systematically
studies *deformations of the representations of concepts*. In that
study, the aesthetic impulse initially occurs as a detonator,
as a buttress (in Peircean firstness) for the elaboration of
a vague image or conjecture, which the mathematician
then submits to meticulous contrasts, through delimitations, definitions, examples and theorems. In turn, in that
contrasting (submitted to forms of Peircean secondness
and thirdness), the aesthetic impulse occurs as a regulator,
as a functor of equilibrium, symmetry, elegance, simplicity. There is a perpetual *double circulation* of aesthetic and
technical factors in the creative act in mathematics. But
Valéry goes on to underscore the central importance of
deformations, displacements and tensions in imagination. We
have, for our part, abundantly emphasized the presence
of these movements in contemporary mathematics, whose
imaginative expressions are thus forged almost *symbioti-
cally* with the most demanding forms (*'SPECIAL efforts'*) of
the imagination, following Valéry.

In mathematical practice itself we can observe the
force of certain aesthetic tensions that, even if they do not
govern the discipline, at least determine the *climate* of some
of its fragments – counterpointing certain foundational
branches outlined in Hilbert's tree, certain impressions

---

403 Ibid., 220 (Valéry's italics and small caps).

of correlative density in Gromov's 'clouds' [263]. There exists a sort of *aesthetic meteorology*, hiding behind the variability of many mathematical phenomena. Gromov's *cloudiness* is typical of contemporary mathematics, and seems to be incomprehensible – or worse, invisible – from elementary, or even modern, perspectives. The shattered tree of Hilbert (complexity, undecidability) is a tree by turns displaced, deformed, and stretched, with extraordinarily dense nodes (the explosion and penetration of complex analysis in the most unexpected domains of mathematics, for example), whose expansive force – *detonating and regulating* – becomes a new object of study. The visions of 'cohomologies everywhere' in Grothendieck [146], of 'groups everywhere' in Zilber [256], or 'metrics everywhere' in Gromov [259], ultimately answer to a new aesthetic sensibility, open to contemplating the local variations of (quasi-)objects through global environments of information transformation. The aesthetic regulation that allows the invasion of cohomologies, groups or metrics be calibrated is decisive.

Turning to particular cases, we can observe a few complex overlappings between aesthetics and technique in contemporary mathematics. Many of the examples *combine* a sort of romantic detonator (recall Langlands's exclamation about the '*romantic side* of mathematics' [185]) and a regulatory transmodern scaffolding. The *interlacing of romanticism and transmodernity* is perhaps surprising at first, but less so when we notice that many great romantics – Novalis, Schelling and Goethe in particular – have taken up, with technical instruments far less robust than the contemporary ones, extensive studies of the *trans-*. To start with, two of the major programs of contemporary mathematics, Grothendieck's motivic cohomology and the Langlands correspondence, explicitly trace themselves back to romantic impulses (Grothendieck's 'grand vision' **[163]**, Langlands's 'mathematics that *make us dream*' **[186]**), and are *combined* with extensive scaffoldings for regulating the transmodern deformations and shiftings in play (EGA **[166]**, the functoriality associated with the Langlands group **[184]**). In a more restricted form, Gromov's *h*-principle **[263]** is the result of an initial, global, romantic intuition (the synthetic penetration of the ideas of holonomy and homotopy into differential realms) and of an extensive subsequent elaboration of hierarchies and local, analytic conditions under which the *h*-principle can be incarnated. The vision of mathematics, according to Lawvere, meshes perfectly with the bipolar tension between romanticism and transmodernity; in his case, the romantic ground corresponds to dialectical intuitions, a gazing into abysses **[190]**, while his reflection's transmodern richness is played out in a multiplicity of new mathematical quasi-objects (in particular, subobject classifiers and sheaves in an elementary topos **[195]**) by which one can gauge fluxions and deformations between opposites. Shelah explicitly (and perhaps polemically) proclaims the *primordial role* of beauty in his understanding of mathematics [200], Zilber plunges into the *abyssal hiatus* of interlacings between model theory and noncommutative geometry [258], Connes underscores the necessity of knowing the *heart* of mathematics [226]: strongly romantic sentiments, potentiated and reoriented by the various migratory modulations of transmodernity.

The 'continuous increase of potentiality', the *summum bonum* of aesthetics according to Peirce,[404] lies behind all of these examples. As we hope to have shown in these pages, contemporary mathematics presents a remarkable *continuous increase of potentiality and reasonability*. In full harmony with the *summum bonum*, the mathematics of the last few decades amplifies our technical, imaginative and rational capacities. A sprawling beauty lies in the work of our great contemporary mathematical creators. A **synthetic** vision allows us to link together apparently distant strata of mathematics and culture, helping us to break down many artificial barriers. Not only can today's mathematics be appreciated through epistemic, ontic, phenomenological and aesthetic modes, but, in turn, it should help to transform **philosophy**. And, as Goethe tells us, we must not forget that 'the most important thing, nevertheless, continues to be the **contemporary**,

---

404 See Barrena, *La razón creativa...*

because it is what most clearly reflects itself in us, and us in it' (our epigraph). This is as true of **mathematics** as of any other realm of human culture. We hope that this *Synthetic Philosophy of Contemporary Mathematics* may help to make up part of the reflection that is demanded – in honor of the human spirit – by one of the most astonishing adventures of thought in our times.

Index of Names

A

Abel, Niels Hendrik 24, 75, 76
Airy, George Biddell 231
Alexander, James Waddell 67, 233
Alunni, Charles 93n
Apéry, Roger 225n
Araki, Huzihiro 221n
Ardao, Arturo 331n
Argand, Jean-Robert 94
Aristotle 89, 329n
Arnold, Vladimir 216n, 335
Artin, Emil 85, 181n, 185n, 227n, 303
Atiyah, Michael 43, 136, 137, 207–213, 215, 221n, 222, 223, 275, 321, 335, 341, 351, 363

B

Badiou, Alain 12n, 71, 88–90, 91, 95, 307n, 312n, 329
Baez, John 98, 125
Bailey, T.N. 184n
Banach, Stefan 31, 135n, 221n, 250, 252
Barrena, Sara 349n, 375n
Barwise, Jon 322n
Batt, Noëlle 93n
Baudelaire, Charles 188n
Baxter, Rodney James 224n
Bayer, Pilar 178n
Beltrami, Eugenio 220
Bénabou, Jean 325n
Benacerraf, Paul 11, 91, 104–105, 279n, 281, 297
Benjamin, Walter 37

Berger, Marcel 260n, 261n
Bernays, Paul 83n
Bernoulli, Jacob 81
Betti, Renato 317n
Birkhoff, Garrett 59, 200n, 253
Blumenberg, Hans 330, 342, 342–344
Bohr, Niels 318–319, 319
Bolyai, János 261n
Bolzano, Bernard 79n, 249, 251n
Bombieri, Enrico 227n
Borel, Armand 42n, 80, 91, 249, 250, 303
Borel, Émile 42n, 80, 91, 249, 250, 303
Botero, Juan José 16
Bouquiaux, Laurence 365n
Bourbaki 53, 76, 77, 96
Boussolas, Nicolas-Isidore 279n
Brentano, Franz 329n
Breuil, Christophe 187n
Brouwer, Luitzen Egbertus Jan 57, 80
Brown, Lawrence Gerald 223
Brown, Ronald 98
Burgess, John 103, 105

C

Caicedo, Xavier 11n, 16, 283n, 307, 362
Canclini, Néstor García 366n
Cantor, Georg 57, 80, 89, 92, 102, 121, 194, 322n
Cardona, Carlos 16
Carlson, James 227n

D

E